FIFTY CLEVER BASTARDS

BY

MARTIN FONE

Copyright © Martin Fone 2016
This book is sold subject to the condition that it shall not, by way of trade or otherwise, be lent, resold, hired out, or otherwise circulated without the publisher's prior consent in any form of binding or cover other than that in which it is published and without a similar condition including this condition being imposed on the subsequent publisher.
The moral right of Martin Fone has been asserted.
ISBN-13: 978-1537047287
ISBN-10: 1537047280

This book would not have been possible without the love and support of my wonderful wife, Jenny aka TOWT. I am eternally in her debt.

This is a work of creative nonfiction. This book has not been created to be specific to any individual's or organizations' situation or needs. Every effort has been made to make this book as accurate as possible. This book should serve only as a general guide and not as the ultimate source of subject information. This book contains information that might be dated and is intended only to educate and entertain. The author shall have no liability or responsibility to any person or entity regarding any loss or damage incurred, or alleged to have incurred, directly or indirectly, by the information contained in this book.

CONTENTS

Introduction *1*
1 Valerian Abakovsky *3*
2 Abu Nasr Isma'il ibn Hammad al-Jawari *6*
3 Harvey Ball *9*
4 James Miranda Stuart Barry *12*
5 Trevor Baylis *15*
6 Laszlo Biro *18*
7 Alexander Aleksandrovich Bogdanov *21*
8 Karlheinz Brandenburg *24*
9 John Romulus Brinkley *27*
10 William Bullock *30*
11 Robert Bunsen *33*
12 Captain Cowper Phipps Coles, C.B., R.N. *36*
13 Karl Drais *39*
14 Douglas Engelbart *42*
15 Guiseppe Marco Fieschi *45*
16 Sieur Freminet *48*
17 Luigi Galvani *51*
18 George Garrett *54*
19 James Goodfellow *57*
20 Charles Goodyear *60*
21 Wilson Greatbatch *63*
22 William Harvey *65*
23 Elias Howe Jr *68*
24 Horace Lawson Hunley *71*
25 Marie Killick *74*
26 Ron Klein *77*
27 Otto Lilienthal *80*
28 Antonio Meucci *83*
29 Thomas Midgley *86*

30 Tom Ogle 89
31 Denis Papin 92
32 Arthur Paul Pedrick 95
33 Anthony E. Pratt 98
34 Louis le Prince 101
35 Robert Recorde 104
36 Sylvester H. Roper 107
37 Jean-François Pilâtre de Rozier 110
38 Jonas Salk 113
39 Carl Wilhelm Scheele 116
40 Walter L. Shaw 119
41 Henry Smolinski 122
42 Percy Spencer 125
43 Hugh Edwin Strickland 128
44 Joseph Swan 131
45 Max Valier 134
46 Edward Vernon 137
47 John Walker 140
48 Henry Winstanley 143
49 The Winstons 146
50 Xerox Corporation 149

I would like to acknowledge the obvious debt I owe to my mother, Brenda, and my mother-in- law, Vera Foord.

INTRODUCTION

I have always been fascinated by inventors. It is a truly wonderful and awe-inspiring gift to be able to spot a problem or an opportunity and apply the grey cells to come up with a solution.

The road to becoming a successful inventor is paved with difficulties and this book is designed to celebrate those who fell by the wayside. Some were killed by their own inventions; some didn't, either through omission, fraud or philanthropic gesture receive the rewards their inventions warranted, and others either took their invention to the grave or came to a mysterious end.

You will find in here inventors of some of the things we take for granted in modern life and, perhaps, you will be surprised to find that they did not receive the financial rewards that their inventions merited. You will find aviators from the first millennium CE, pioneers of medicine and science and individuals who took their zeal to learn just a bit too far.

My simple idea is to shine a spotlight on these individuals and to construct an imaginary Hall of Fame. Each pen picture has been culled from posts in my eclectic blog, windowthroughtime.wordpress.com – feel free to look at it.

And a word about the title. 'Clever bastards' is an English term of endearment to describe someone who is a cut above the rest in terms of intelligence and ingenuity. It is also used pejoratively to describe someone who is a little too clever for their own good. As Ian Dury so memorably said, there ain't half some clever bastards.

Finally, this book would not have been possible without the love and support of my wonderful wife, Jenny aka TOWT. I am eternally in her debt.

1

VALERIAN ABAKOVSKY
(1895 – 1921)

The first inductee to our Hall of Fame is the Latvian inventor, Valerian Abakovsky, whose claim to fame was developing a propeller driven railway car.

A key prerequisite of an inventor is the ability to think outside of the box. In the second decade of the 20th century, the railway was a well-established means of transportation. The problem, though, before the harnessing of diesel and/or electricity as the principal means of power was that the speed of the train was constrained by the amount of power that could be generated by shovelling coal into the engine's boiler. However, aeronautics was now an established science and the engines required to provide the aircraft with the thrust necessary to get off the ground and travel at speed were both more compact and more powerful than the rather antiquated means of locomotion that trains were reliant upon. If you were wanting to develop a high-speed train, why not marry the two technologies?

This was probably the train of thought that our hero followed in the periods that he was hanging around waiting for the Soviet dignitaries to finish doing whatever they were doing and get back into his car. Unlike many a would-be inventor, the Latvian chauffeur actually lobbied to be allowed to put his brain wave into practice, doubtless using the close contact with officialdom that his day job afforded him. You know the sort – the course of least resistance is to give into someone who is always banging on about their pet theory.

Anyway, Abakowsky was allowed to get on with it and produced a prototype which looks like an enclosed boat-like wagon with a dirty great propeller on the front. It was unveiled in July 1921 and initial trials went well. The idea was that it would be used to convey the Soviet top brass at speed to and from Moscow. On July 24 1921, a high-powered delegation, led by the revolutionary and close friend of Joseph Stalin, Fyodor Sergeyev, and our intrepid inventor clambered aboard to travel at speed from Tula to Moscow. The trip passed off without incident and the party was encouraged to repeat the experience in order to get back to Tula.

Unfortunately - and we come to expect an unexpected turn of events and a tragic outcome with our inductees – the Aerowagon, as it was dubbed, derailed at high-speed, killing everyone on board. Still, on the plus side, the idea of train powered by an aircraft engine had been firmly established and the six martyrs to the cause of progress were buried in the Kremlin War Necropolis which since 1917 had been the final resting place of the heroes of the October Revolution.

Our hero's legacy lived on. Franz Kruckenberg developed the Schienenzeppelin which was built as a prototype in 1930 and in trials in 1931, this train powered by an aircraft propeller reached speeds of up to 200 kilometres an hour. Alas, the build up to World War 2 consigned it to the scrapheap.

Closer to home the Scottish engineer, George Bennie, built a prototype track and railcar for his Bennie Railplane at Milngavie. Typically, whilst everyone thought it was a brilliant idea, no one came up with the dosh to turn it into reality.

Valerian, for your ingenuity and self-sacrifice, you are a worthy inductee.

2

ABU NASR ISMA'IL IBN HAMMAD AL-JAWARI
(Died 1002 or 1008)

Our second inductee is al-Jawari, the noted lexicographer who died around 1002 or 1008 CE. Hailing from what is now Kazakhstan, Al-Jawari's main claim to fame, although not why he receives this nomination, is his compilation of al-Sihah, a lexicon which contains some 40,000 entries.

His light-bulb moment was to put the entries into an alphabetical order in which the last letter of a word's root is the main criterion by which the order is established. Although it was incomplete at the time of his death – it is said that a student completed the magnum opus – it stood the test of time, becoming one of the main Arabic dictionaries in the medieval era. Many of its entries became the basis for an Arabian to Turkish dictionary which was the first book to be published in the Ottoman empire on a printing press (in 1729).

The urge to fly must have been a primeval instinct amongst man. After all, the birds are so free and can travel great distances unhindered by the obstacles we find on land. The legend of Daedalus and Icarus testifies to the antiquity of the desire and, indeed, of its perils. We tend to think – or at least Occidentals do – that experimentation with flight is a fairly recent phenomenon.

Think again and consider the derring-do of Abbas ibn Firnas (810 – 887CE), a polymath based in Cordoba. Writing some seven centuries later, a Moroccan historian, al-Maqqari, comments that among Firnas' curious experiments, was one where he covered his body with feathers, attached a couple of wings to his body, climbed up high and launched himself into the air. According to what al-Maqqari considers to be trustworthy writers, he flew a considerable distance but *"in alighting again at the place whence he started, his back was very much hurt"* because he had forgotten to provide himself with a tail.

And then closer to home we have Eilmer, a Benedictine monk at Malmesbury Abbey at the turn of the 12th century. In Gesta Regum Anglorum, written by a fellow monk, William of Malmesbury around 1125, we learn that Eilmer fixed wings to his hands and feet and launched himself off the tower of Malmesbury

Abbey. Remarkably, if William is to be believed, he flew more than a furlong before landing proved his undoing. "*Agitated by the violence of the wind and the swirling of the air, as well as by awareness of his rash attempt, he fell, broke both his legs and was lame ever after*". Eilmer, too, attributed his failure to forgetting about giving himself a tail.

Still, Firnas and Eilmer should count their blessings or give thanks to their respective Gods that they only suffered debilitating injuries as a result of their attempt to follow in the flapping wings of Daedalus and Icarus. As you might expect, what earns al-Jawari his place in our Hall of Fame over and above the strong claims of the other two is that his folly brought about his demise.

It is thought in an attempt to emulate Firnas and, presumably, to add a bit of spice to his otherwise dull but laudable work as a lexicographer, al-Jawari climbed on to the roof of a mosque in Nishapur wearing the obligatory wings – as for a tail, my researches have failed me. Inevitably, too, after launching himself into the air, he plunged to the ground, killing himself in the process.

Al-Jawari, as the representative of the aviators of the first millennium, you are a worthy inductee to our Hall of Fame.

3

HARVEY BALL
(1921 – 2001)

The emoji is considered to be the fastest growing language. For me the first manifestation of man's desire to return to pictograms was the smiley face, generally yellow with an inane smile, which was so ubiquitous that, rather like traffic cones, you wonder what brought them about and who was it that first devised them. Wonder no more because this is where the latest inductee to our illustrious Hall of Fame, American commercial artist Harvey Ball, comes in.

The insurance industry can rarely be accused of being a force for good or improving human existence but this might be an exception. The State Mutual Life Assurance Company had bought the Guarantee Mutual Company and as is often the way in these circumstances, was facing a staff morale issue as a result of trying to integrate the two organisations. Someone in the higher echelons, showing the genius that got them there, decided that what would persuade the reluctant employees to do twice the work for the same pay was a feel-good campaign.

Ball was employed in 1963 to design an image that could be used on buttons, desk cards and posters. Within less than ten minutes – you wonder what took him so long, so rudimentary is the design – the smiley face was born. And it was an instant success, achieving State Mutual's objective of getting its staff to smile while going about their everyday tasks of answering the telephone, turning down claims and making colleagues redundant. It was a roaring success.

Great news, you would think, for our Harvey but alas he didn't share in the success of the image that he created for the simple reason that he did not copyright or patent the design. All he got directly from his endeavours was his fee from State Mutual of $45. If it makes you feel any better, State Mutual didn't benefit either because they didn't patent or copyright the image.

Where there is an opportunity, though, others will step in. Step forward brothers Bernard and Murray Spain from Philadelphia. They came across Ball's badge design and thought that it, together with the slogan, "Have a happy day" would go a long way to curing America's post-Vietnam blues. And so they copyrighted the icon and slogan and started producing their own novelty items. By the end of 1971 they had sold over 50 million badges and other bric-a-brac, generating a substantial profit. They changed the slogan to "Have a nice day" along the way and it has become a popular – and irritating, in my view – alternative to "Goodbye".

In France in 1972, journalist Franklin Loufrani became the first person to register the mark for commercial use when he used a smiley face to highlight the, admittedly rare, instances of a good news story in the pages of France Soir. He subsequently trademarked the image, known as Smiley, in over 100 countries and launched the Smiley Company by selling T-shirt transfers. Today the Smiley Company is one of the top 100 licensing companies of the world.

Loufrani's son poured *eau froide* over the suggestion that Ball was the originator of the design. Loufrani suggests that the image is so basic that it cannot be assigned to anyone, pointing out on the company website the remarkable resemblance of a stone age carving dating from 2,500 BCE found in a French cave to the smiley.

Ball showed remarkable sangfroid when he saw the appropriation of and commercialisation of his image. He shrugged his shoulders and said, *"Hey, I can only eat one steak at a time, drive one car at a time"*. For this stoicism Harvey, you are a worthy inductee to our Hall of Fame.

4

JAMES MIRANDA STUART BARRY
(1789/99 – 1865)

The latest inductee into our Hall of Fame is James Barry who had a career as a military surgeon in the British Army. As you can see from his dates, his early life is shrouded in mystery, more of which later.

Graduating from the University of Edinburgh Medical School in 1812, he successfully passed his exams at the Royal College of Surgeons in England in July 1813 and was commissioned as a Hospital Assistant with the army where he quickly rose up the ranks to become an Assistant Staff Surgeon. It is thought he saw service at the Battle of Waterloo.

One of the benefits of having a scalpel and being in the British Army was that it gave one the opportunity to see the world. Barry served in India and then arrived in Cape Town. Within a couple of weeks of landing there he had become Medical Inspector for the colony and began to campaign for and implement a better water system. It was there that Barry achieved his first claim to fame – he performed the first known successful Caesarean section with both mother and baby – named James Barry Munnick – surviving the ordeal.

But Barry was clearly a prickly character, not suffering fools gladly, because wherever he went around the Empire, he earned the enmity of his fellow Brits and often had to leave quickly. He was said to have had a good bedside manner, was a vegetarian and tee-total (we can't all be perfect) and recommended wine baths for his patients. By 1857 he had become an Inspector-General of Hospitals in Canada and campaigned for better food, sanitation and proper medical care for prisoners and lepers as well as soldiers and their families. He campaigned for and introduced pears into the military diet, immediately improving the soldiers' daily intake of Vitamin C. Retiring in 1864 – reputedly against his wishes – he returned to England and died the next year from dysentery.

It was his death which unravelled Barry's secret and why this medical campaigner deserves his elevation to our hallowed ranks. Sophia Bishop, a charwoman, took care of his body and when she examined Barry's body she was astonished to find that he was a female. According to a Major McKinnon, "*Bishop had examined the body and was a perfect female and farther that there were marks of him having had a child when very young*". McKinnon concluded that Barry was a Hermaphrodite.

Whether that was the case or not, it seems that Barry was identified or assigned as a female at birth and named Margaret Ann Bulkley, raised as a girl but, because the times were not as liberal in those days as now, chose to live as a man so that he could be accepted to a university and study to be a doctor. It is astonishing that (s)he pulled it off and there is evidence that many in the know were complicit in the deception.

For all Barry's failings, he was a medical campaigner and deserves recognition as the first British woman to become a qualified medical doctor.

James Barry, as British medicine's foremost gender-bender, you are a worthy inductee.

5

TREVOR BAYLIS
(1937 – Present)

The radio is a wonderful invention. It provides company to the lonely, disseminates information and forms of entertainment and allows nations to talk to nations. The receiver, though, needs some form of power, typically electricity or via a battery, to work and without it you are snookered. This is where the latest inductee into our illustrious Hall of Fame, Trevor Baylis, comes in.

With AIDS sweeping through Africa in the 1990s, the radio was an important medium for spreading the vital health information to the more remote communities. The problem, though, was that batteries needed to power the radios were hard to come by and expensive. A TV programme on the subject aired in 1993 spurred Baylis into action.

Tinkering around in his workshop he began experimenting with a hand brace, an electric motor and a small radio. He found that the brace turning the motor acted as a generator which could produce enough energy to power the radio. His next adaptation was to add a clockwork mechanism that meant that a spring could be wound up and as it unwound, the radio would play. On winding his prototype for a couple of minutes, the radio played for 14 minutes. Baylis had invented the windup radio which, as it was totally reliant upon human power, could be a godsend for so-called underdeveloped countries.

Baylis tried to interest companies in his invention but those he approached were dubious as to its commercial value. Nonetheless the invention was showcased on an edition of "Tomorrow's World" in April 1994 and piqued the interest of an accountant, Christopher Staines, and South African entrepreneur, Rory Stear. With funding from the Liberty Life Group, a South African insurer, Stear and Staines set up a company, BayGen Power Industries, in Cape Town to make a commercial version of Baylis' invention.

Although Baylis had patented his invention and was involved in the BayGen business, as you would expect with an inductee, all did not go to plan. The company made a small but important change to the original design, using the spring to charge a battery rather than generating the power directly. This subtle change took the radio outside of Baylis' patent and so he lost control of the product and the revenues that followed from it.

And the radio was successful. A second generation radio was developed in 1997 aimed at Western consumers which would run for an hour on a thirty-second wind. We have one which we use in the garden. The range of products using the technology grew to include a torch – very useful in power cuts – a mobile phone charger and a MP3 player. Showered with honours – he was awarded an OBE in 1997 and CBE in 2015 and was so regularly in the media that he became Pipe Smoker of the Year in 1999 – Baylis saw very little of the financial benefits that flowed from his inventions.

In his latter years, Baylis has devoted part of his time to lobbying the British Government to better protect the rights of inventors so that they do not suffer the same fate as him. As he said, *"I was very foolish. I didn't protect my product properly and allowed other people to take my product away. It is too easy to rip off other people's ideas"*.

Nonetheless, Trevor, for bringing the world the wind up radio and getting ripped off in the process, you are a worthy inductee into our Hall of Fame.

6

LASZLO BIRO
(1899 – 1985)

Cursive handwriting is now almost a dead art, alas, but back in the days before computer keyboards the development of a writing tool which was quick and clean was a godsend. Although the fountain pen had been a distinct leap forward from the quill feather, it was still a messy procedure. Even writing the simplest of documents you ran the risk of inky stains on your fingers or smudge marks on the page or both. Surely a fortune would be in store for someone who invented a cleaner more efficient form of writing instrument?

So, at least, thought the latest inductee to our Hall of Fame, Laszlo Biro, a Hungarian Jew by origin. As well as being a bit of an inventor in his spare time – Biro held patents for a steam-powered washing machine and an automatic transmission for a car – his day job was that of a newspaper reporter. He noticed that ink from inkwells or in fountain pens took several minutes to dry and smudged easily whereas the ink used for typesetting a newspaper dried much more quickly.

So Biro started out by seeing whether he could combine the newspaper ink with fountain pen technology. The problem he encountered was significant – to work, ink has to flow from the tip to the paper but the press ink was too thick to flow. What he needed was a way to transfer the quick-drying ink to the page without requiring the ink to flow. The answer, thought Biro, was to close the end of the pen instead of affixing a nib, leave an opening with just enough room for a tiny metal ball to spin against the reservoir of ink and distribute it on to the paper. Simple.

Biro's brother, Gyorgy, was a chemist and he helped in refining the ink and this in conjunction with a rolling ball in a socket at the pen's tip worked a treat. Smudges, blotches and inky fingers were soon to be a thing of the past. So successful was the design that it has remained virtually unchanged ever since.

But that was not the end of the story. The first problem for Biro was his Semitic ancestry in 1930s Hungary. He and his family were forced to flee, arriving in Paris where in 1938 he successfully applied for a patent and then fled to Argentina. He was soon able to hold patents in Argentina, Hungary,

Switzerland and the States. The pen went into commercial production in 1944 but was not immediately successful, mainly because the pen relied upon gravity to push the ink to the ball so unless you held it perfectly upright it would not work. Brother Gyorgy fixed the problem by employing a capillary action to pump the ink to the tip. One major still remained – its propensity to leak and ruin clothing.

Biro's breakthrough came courtesy of the RAF that saw its merits for navigators who had to mark points on maps quickly and permanently and they placed the first bulk order, for 30,000. But competitors were quick to spot the potential of Biro's invention, notably a Chicago businessman called Milton Reynolds who was able to copy most of Biro's design into Reynold's Rocket. Sold exclusively through a New York department store, Gimbels, it was an overnight sensation.

The revolution in the ball point pen occurred in the early 1950s when Marcel Bich resolved all the problems associated with the ball point by creating the Bic Crystal, an efficient, effective, cheap and throwaway version of Biro's original concept. It was a rip-roaring success and around the world some 57 pens a second are bought. But Biro didn't share in its success, although he gave his name to the writing instrument.

Laszlo Biro, for inventing the ball point pen and not profiting from it, you are a worthy inductee into out Hall of Fame.

7

ALEXANDER ALEKSANDROVICH BOGDANOV (1873 – 1928)

The latest inductee into our hall of fame is the Russian physician, philosopher, science fiction writer and revolutionary, Alexander Bogdanov.

A Belarusian by origin, Bogdanov was a leading figure in the Bolshevik faction of the Russian Social Democratic Labour Party, being a co-founder. He became a rival to Lenin and once he had lost out to him he was expelled from the party in 1909. However, following the Russian revolution he became an influential opponent of the regime adopting a more purist Marxist position.

A doctor and psychiatrist by training, Bogdanov became interested in what is now known as systems theory and synergetics. He developed Tektology which was an attempt to unify all social, biological and physical sciences by considering them as a system of relationships. He was interested in trying to find the underlying organisational principles – a real life manifestation, perhaps, of what George Eliot's Dr Casaubon was trying to do for Greek mythology.

The reason why Bogdanov warrants a place in our illustrious pantheon of misguided geniuses is because of his interest in eternal youth – on some days for people of my vintage, this prospect can seem appealing. In 1924 he started a series of experiments involving blood transfusions in an attempt to reach this state of nirvana or, at least, to achieve some degree of partial rejuvenation. Ironically, the sister of his great political rival, Lenin, actually volunteered to participate in the deluded scientist's experiments. After eleven transfusions, Bogdanov reported a number of positive developments including that his eyesight had improved and that his developing baldness had stopped. One of his revolutionary comrades reported that Bogdanov was looking 7 to 10 years younger after one particular session. Spurred on by his success he went on to found the Institute for Haematology and Blood Transfusions in 1925/6.

However, his experiments into blood transfusions caused his untimely demise. In 1928 he took the blood of a student who was suffering from malaria and tuberculosis and the combination of the diseases introduced into his system and blood type incompatibility did it for him. Ironically, the student who received Bogdanov's blood made a complete recovery.

A truly worthy candidate for induction.

8

KARLHEINZ BRANDENBURG
(1954 – Present)

You remember those carefree, innocent days when your only access to music whilst on the hoof was via your own vocal chords? And then came along the cassette tape and the Walkman which for all its imperfections got us into the groove of listening to music on the move. And then came the MP3 format which revolutionised it all. But who invented the format and what has Suzanne Vega's opening track on her 1987 album Solitude Standing, Tom's Diner, got to do with it? The latest inductee to our illustrious Hall of Fame holds the answers.

Born in Erlangen in 1954 Brandenburg was a PhD student and audio engineer who joined the Moving Pictures Expert Group, an international collaboration of scientists founded in 1988, who were trying to develop an efficient compression tool which would compress full-motion video and high-quality audio into digital form.

The group achieved some early success, developing a large machine the size of a refrigerator which compressed a sound file to just 8% of its original size. Clearly this wasn't really a practical option and what was needed was an algorithm which squeezes a sound file into the smallest possible file. And this is where our inductee came in.

His PhD study director suggested Brandenburg should look at how to transmit music over an ISDN telephone line in a way that maintained the quality of the sound and overall experience for the listener. The piece of music he chose to experiment on was Vega's Tom's Diner which, if you are familiar with the track, is an a cappella version and which meant that the algorithm had to be precise in what it retained and what it discarded. Eventually Brandenburg cracked it and the algorithm which made the MP3 format viable by shrinking the data while retaining the audio quality to a level that was acceptable to most listeners, pace Neil Young, was born.

You would think that such a discovery which has been pretty universally adopted would be the key which unlocked untold riches for Brandenburg. But, alas, it was not the case and that is the reason why our inventor has been inducted into our Hall of Fame. Following the successful completion of his doctorate,

Brandenburg joined the Fraunhofer Society, one of Germany's major research institutions, and with his colleagues tried to work out what to do with the audio compression software.

Lack of funds for marketing meant that they released l3enc, the code necessary to translate .wav files into MP3 format, as shareware in July 1994. There was a modest charge for using it, around $250, and this was followed up in September 1995 with Fraunhofer WinPlay3, the first real-time MP3 software player, allowing people to store and play MP3 files on their PC.

But Brandenburg and Fraunhofer didn't have the field to themselves. Other teams were working on ways to crack the problem and there were a number of competing software solutions, all vying to be adopted as the international standard. MP3, however, proved to be the most and efficient format as did AAC, another Fraunhofer development, which was adopted as the equivalent standard for iPods.

The inventions generated millions in royalty payments for the non-profit Fraunhofer. As for Brandenburg, he didn't achieve mega-riches but at least German law allows researchers to a share in the profits of their inventions.

9

JOHN ROMULUS BRINKLEY
(1885 – 1942)

Without doubt, Viagra has been a boon for certain sectors of the male population. Prior to its development, males were prone to exploitation by many a quacksalver in their desperation to find a cure for their loss of potency.

Step forward our latest inductee into our Hall of Fame, John Brinkley.

Starting out his medical career as a snake-oil salesman, Brinkley actually gained some medical qualifications and developed an air of respectability through setting up a 16-room clinic in Milford where he was successful in treating and restoring back to health many who had succumbed to the virulent and deadly outbreak of influenza in 1918.

He then plied his fevered mind to the problems of male impotency. The story goes that when he was approached by someone to fix a patient's impotency problems, he jokingly retorted that the unfortunate patient would have no problems if he had a couple of buck goat glands implanted in him. The patient took him up on his word and Brinkley was persuaded to transplant the goat glands into him. Surprisingly, the operation was successful and nine months later the patient had sired a son, named unsurprisingly, Billy.

This success spurred our mad scientist on and he started advertising and was soon carrying out operations to transplant the testicular glands of goats into his male patients at $750 a shot. Over 16,000 men had their scrotums cut open and goat gonads placed in their sacs. The patient's body would absorb the gonads as foreign matter; the organs were never accepted as part of the body. At best, the men's bodies simply broke down the goat tissues and healed up but some were not so lucky. Their fate was compounded by Brinkley's questionable medical training and his custom of performing many of his operations whilst the worse for drink.

It is thought that 43 died directly as a result of his operation whilst hundreds more died from complications such as gangrene and infection. Not unsurprisingly, the authorities in Kansas revoked his medical licence in 1930 but that did not deter our mad clinician. He skipped over the border to Mexico and started up again performing his goat-gland operations.

Justice eventually caught up with him though. Morris Fishbein made a career out of exposing medical frauds and in 1938 published "Modern Medical Charlatans" which included a damning indictment of Brinkley's career and methods. Brinkley responded by suing Fishbein for $250,000. The jury found in Fishbein's favour and this verdict led to a barrage of lawsuits against Brinkley, claiming in total around $3m.

Our hero then had the tax authorities round his neck and, inevitably, he declared himself bankrupt in 1941. To compound his problems, the US Post Office Department began investigating him for mail fraud and whilst these investigations were in train Brinkley suffered three heart attacks, had a leg amputated as a result of poor circulation and finally succumbed to heart failure, dying in San Antonio penniless.

Brinkley was quite a character. He pioneered radio advertising and was a proponent of the border blaster radio phenomenon – illegal radio stations targeting the US from other countries, usually Mexico. He also ran for governor of Kansas twice.

A truly worthy inductee for our Hall of Fame.

10

WILLIAM BULLOCK
(1813 – 1867)

The latest inductee to our Hall of Fame is William Bullock who transformed the printing industry with his development of the rotary perfecting press.

Bullock was born in Greenville, New York and was orphaned at an early age. He started his working career early out of necessity as a machinist and iron founder and through self-education developed an interest and profound knowledge of mechanics. By the mid-1830s he was running his own machinery shop in Savannah, Georgia and developed a shingle-cutting machine. Unfortunately, his business went bust because he was unable to market his invention.

Moving back to New York – he had found time to sire 13 children with two wives, in the interim – he made artificial legs. But soon he was able to give full rein to his inventive side, designing, amongst other things, a cotton and hay press, a seed planter, a lathe cutting machine and a grain drill. It was this latter invention that brought him to national prominence when he was awarded second prize by the Franklin Institute in 1849.

Shortly afterwards, our hero moved into the newspaper industry editing the American Eagle and then building a high-speed press for the nationally circulated Leslie's Weekly in 1860. The newspaper industry was growing like Topsy – by the 1850s there were over 2,500 newspapers in the US alone – and the technological deficiencies of the printing presses were seriously inhibiting production capabilities. As early as 1835, Sir Rowland Hill – he of postage stamp fame here in Blighty – suggested that the next major step forward should be the development of a press that was capable of printing on both sides of the paper at the same time. It was to this problem that Bullock applied his mind.

By 1861 he had cracked it with the rotary perfecting press and he was awarded a patent two years later, after the prototype had been installed at the Cincinnati Times. In 1865 the Philadelphia Inquirer had installed the first fully functioning model. Bullock's machine contained a number of technological breakthroughs – it allowed for continuous large rolls of paper to be fed automatically through its rollers, thus eliminating the necessity to hand-feed paper into the machine; the press was self-adjusting, allowed printing on both sides of the paper, folded the paper and with a sharp serrated knife which rarely needed sharpening, cut the sheets with rapid precision. The combination of all these features revolutionised the efficiency of the printing process. His early models were able to produce 12,000 sheets an hour and later, after further refinements, notched up an impressive 30,000 sheets an hour. His design is still used today.

But sheer genius is not enough to make it into our Hall of Fame - there has to be a flaw in your character or you have to suffer a monumental stroke of ill-fortune. It was the latter that earns Bullock his elevation to our rolls. On April 3rd 1867, our hero was making some adjustments to one of his new presses being installed for the Philadelphia Public Ledger newspaper. He tried to kick a driving belt onto a pulley but, unfortunately, his leg got caught in the machinery and was crushed. Within a few days he had developed gangrene and on 12th April died on the operating theatre when surgeons were trying to amputate his leg.

Developing a revolutionary piece of technology which was the cause of his death – William Bullock is a truly worthy inductee to our Hall of Fame.

11

ROBERT BUNSEN
(1811 – 1899)

I regard it as a badge of honour – I managed to get a creditable grade in my Chemistry O level without ever once lighting a Bunsen burner. The latest inductee into our Hall of Fame, the German chemist, Robert Bunsen, developed the burner to which he gave his name and which has enabled schoolchildren down the ages to experiment with chemical reactions, both officially and, more enjoyably, unofficially.

Without doubt our Robert was a clever bastard. Born in Gottingen in 1811, which at the time was in the short-lived independent kingdom of Westphalia but was soon absorbed into the kingdom of Hannover, Bunsen studied chemistry, mineralogy and mathematics at the local university. After graduating he taught at the local university and started to conduct experiments into the solubility or otherwise of arsenious acids. His discovery that using iron hydrate oxide as a precipitating agent to produce an antidote to arsenic poisoning – a major problem in those days as arsenic was used to produce the green colouration in wallpapers and was responsible for many household deaths – is still used today and is considered to be the most effective antidote.

In 1841 Bunsen created the Bunsen cell battery – it is amazing how throughout his career he discovered things which bore his name – which used the cheaper carbon electrode instead of the platinum electrode, a development which reduced the cost of producing a battery.

Moving to Heidelberg in 1852, our hero pioneered the development and use of electrolysis to isolate metals such as chromium, magnesium, aluminium, manganese, sodium, barium, calcium and lithium. In 1859 he started to study the emission spectra of heated particles, a branch of scientific enquiry called spectrum analysis. For this work he needed to generate a powerful and intense flame to heat up the particles. After much experimentation, he and his assistant Peter Desaga came up with a burner which threw out a clean and very hot flame – what we now know as the Bunsen burner.

This wasn't the end of his ingenuity. Later on in 1859, perhaps his *annus mirabilis*, he developed what we now know as a prototype spectroscope which enabled him to identify the characteristic spectra of sodium, lithium and potassium. It was through this research that he was able to discover and isolate previously undetected elements, principally caesium and rubidium.

But, I hear you cry, the common characteristic of our inductees is some element of misfortune and all you have told us about Bunsen suggests that he led a life free from misfortune. Not a bit of it! Arsenic, particularly because of the presence of cacodyl which is extremely toxic and undergoes spontaneous combustion when in contact with dry air, was notoriously difficult to work with. Constant exposure to arsenic meant that Bunsen almost died from the build-up of arsenic poison in his system. Worse was to follow in 1843. Whilst experimenting, some cacodyl cyanide blew up in his face causing him to lose his right eye.

Robert, for the sacrifices you made to further our understanding of and protection from arsenic and for inventing, amongst other things, the Bunsen burner, you are a worthy inductee.

12

CAPTAIN COWPER PHIPPS COLES, C.B., R.N.
(1819 – 1870)

The latest inductee to our Hall of Fame is the British sea-captain, Cowper Phipps Coles. Incredible as it may seem to us today, our hero joined the Royal Navy at the tender age of eleven and distinguished himself during the Crimean war at the siege of Sevastopol and by August 1856 had become the commander of a Black Sea paddle steamer, the Stromboli.

It was during this time that Cowper took his first tentative steps into marine craft design. He together with some colleagues constructed a 45 foot raft called the Lady Nancy from twenty nine casks lashed together with spars. The raft was able to carry a 32 pounder gun and because of the small draft of the vessel it was able to navigate the shallows, get closer to the Russians and maximise the damage it could inflict on them. His superior, Admiral Lyons, recognised the strategic edge that such a design provided and sent Coles to the Admiralty to present his ideas. Work commenced on developing a larger raft but, alas, the war ended before construction was complete.

Following the end of the war, on half pay from the navy, Coles set about designing turret towers for gunships. Whilst battle ships were bristling with armaments, they were fixed. On 10th March 1859 our hero applied for a patent for a revolving turret which would revolutionise the effectiveness of battle ships, although the Americans were the first to incorporate revolving turrets into their ships.

Coles then sought to incorporate his two innovations – a vessel with a low draft and a revolving turret – into one design. Coles submitted a number of proposals but met with scepticism from the Admiralty and he had to fight hard, including running what would now be termed a PR campaign enlisting support from Prince Albert amongst others, to get his ideas adopted.

His greatest achievement was to get his plans for the HMS Captain approved. The vessel was designed such that the distance between the deck and the water line was just 8 feet but because of mistakes committed during its construction which made it heavier than envisaged, it actually floated some 14 inches lower. The vessel performed well in trials, being marginally slower than the fastest then ship in the fleet under steam and faster under sail. It weathered a gale successfully.

Alas, as you would expect with our inductees, disaster was soon to strike. On 6th September 1870 the Captain, with Coles on board, was cruising off Cape Finisterre. The wind got up and water started washing over the weather deck, the vessel being so low in the water. Shortly after midnight, the vessel was keeling over at eighteen degrees and was felt to lurch to starboard twice. Before the Captain's orders to drop the foresail could be carried out, the vessel lurched alarmingly and capsized and sank. Just 18 of the crew survived, around 480 perishing, including Coles.

The subsequent inquiry, which was in the form of a court-martial, established that the vessel was inherently unstable, compared with other designs, if the roll was greater than 18°.

Captain Coles, for your innovations in ship design and for paying for your ingenuity with your life, you are a worthy inductee.

13

KARL DRAIS
(1785 – 1851)

Geoffrey Parker's magisterial Global Crisis demonstrated, at great length, the impact of climate change on the political fortunes of the world during the 17th century. It seems that adverse climatic conditions have been the catalyst for ground-breaking inventions in other eras too. This is where the latest inductee to our Hall of Fame, Karl Drais, steps in.

The second decade of the 19th century was also a period of great climatic change – 1816 was dubbed the year without a summer after the eruption of Mount Tambora in Indonesia in 1815 – the largest recorded volcanic eruption – thrust volcanic ash into the atmosphere, causing crops to fail and animals to starve as far west as Western Europe.

In those days, man had very few options available to him to travel around the place. Of course, he could rely on Shanks's pony. Otherwise he was reliant upon the horse, with or without a cart. But the adverse climatic conditions meant that there was little grazing pasture for the horses and they starved in numbers. It was this problem – how to develop an alternative means of transportation – that exercised our hero's mind.

And as we would come to expect from an inductee, he cracked the problem, coming up with what he called the Laufmaschine or running machine. It was a two-wheeled vehicle with both wheels in a line propelled by the rider pushing their feet along the ground as if they were walking or running. The front bar and handlebar assembly was hinged to allow the machine to be steered. His first public outing on the contraption took place on June 12th 1817 when he set out from the centre of Mannheim to a coaching inn in Rheinau. His second trip was from Gernsbach to Baden-Baden.

In 1818 Drais was awarded a grand-ducal privilege to exploit his invention but as Baden had no patent laws, others quickly saw the opportunity and exploited the results of his labours. The machine became popular in France, where it was known as the *draisienne* and in England where it was known as the dandy horse and several manufacturers sprang up there in 1819.

Of course, not everyone welcomed this new road to freedom which had opened up – at least for the well-to-do. Riders preferred to operate their machines on the pavements which offered a slightly smoother ride than the pot-holed roads but this meant that they became a menace to pedestrians and some authorities prohibited their use!

As is the way with inductees, misfortune dogged Drais. He was forced into exile in the 1820s because of political unrest and although he was eventually able to return, he seemed to be cursed. Following the revolutionary uprising in Baden in early 1848 – in a fit of revolutionary fervour Drais renounced his title and styled himself Citizen Karl Drais - the Prussians stepped in and crushed it in 1849, taking Drais' pension amongst others as reparation for the costs of suppressing the revolution.

Drais never recovered and so the father of cycling died penniless in Karlsruhe in 1851. Karl, you are truly a worthy inductee.

14

DOUGLAS ENGELBART
(1925 – 2013)

The computer mouse - now there's a handy little gadget which when plugged into the laptop allows the user, particularly one with poor manual dexterity and/or fingers like tree trunks, to navigate around the screen with ease. The man who developed this simple but ingenious gadget is the latest inductee to our illustrious Hall of Fame, American scientist Douglas Engelbart.

Engelbart, then a Director at the Stanford Research Institute, was working on a project aimed at augmenting the human intellect, pioneering the use of graphical user interface and looking for a way to enable the user to more easily interact with the information being displayed on the screen. The first prototype of what we would call today a mouse was developed in 1964. It had a cord at the front – it was later moved to the rear of the device to get it out of the way – and was a simple mechanical device with two perpendicularly mounted discs on the bottom. By tilting or rocking the mouse you could draw perfectly straight vertical and horizontal lines on the screen.

On behalf of the Research Institute, Engelbart applied for a patent for what was a wooden shell with two metal wheels or in the language of the patent application "*an X-Y position indicator for a display system*". The patent was granted in 1970 and the gadget went by the name of a mouse because of its relatively small body and long tail, the cord coming out of the back. Interestingly, Engelbart was not able to apply for patents for his early versions of windows and GUI because at the time, software was not thought to be capable of being patented.

In 1967 Engelbart experimented with a mouse that was controlled by the foot on the basis that the knee offered better control at slight movements in all directions and in tests it outperformed hand-controlled mice by a small margin. But the knee-trembler was dropped and the prototype of the advanced computer technologies which Engelbart and his team demonstrated on December 9th 1968 featured a hand-controlled mouse. Some of the other technologies on display included video-conferencing, teleconferencing, email, hypertext, word processing, hypermedia, object addressing, dynamic file linking, bootstrapping and a real-time editor – all the bits and pieces within a PC we take for granted.

Perhaps because he was an academic and the mouse was viewed as just an adjunct to the greater project of grappling with assisting the human intellect, the commercial possibilities offered by the mouse were cheerfully ignored. The patent, in fact, seemed to pass through a couple of hands until it came to the attention of one Steve Jobs – yes, him. Never one to overlook a gift horse in the mouth, Jobs set about streamlining the device and marketing it as an aid to facilitate the use of laptop devices.

When accused of stealing and ripping off the idea, Jobs is supposed to have echoed Picasso's famous quote, "*Good artists copy, great artists steal*" and told a journalist that the previous holders of the patent had no clue of the value of the patent at the time. Jobs went on to make his billions but the rewards that should have come Engelbart's way eluded him, qualifying him for elevation to our role of honour. At least when the mouse took off Engelbart's contribution was recognised, if not financially.

Douglas, for your development of the computer mouse for which you received not a bean, you are a worthy inductee into our Hall of Fame.

15

GUISEPPE MARCO FIESCHI
(1790 – 1836)

I have been reading a lot about assassination plots in the 19th century and what has struck me has been how unreliable the weaponry was at the time. Usually indigent, the would-be assassins were reliant upon cheap and inefficient weaponry that generally couldn't hit a barn door and were so difficult to re-load you were left with a single shot.

Our next inductee into the hall of misguided geniuses, Fieschi, born in 1790 in Murato in Corsica, applied his grey cells to the problem. He had been imprisoned for 10 years for theft and forgery – he regarded himself as a victim of injustice – and decided to take his revenge on society by assassinating King Louis Philippe of France. With two accomplices, Morey and Pepin, he devised what he thought would be the ultimate assassination machine, the machine infernale. Recognising that he would have only one shot at his target, Fieschi took twenty guns and fused them together in a way that they could be fired simultaneously. In that way, he was bound to hit his target.

On 28th July 1835, Louis Philippe was passing along the Boulevard du Temple in Paris with his three sons and numerous staff. Fieschi fired his machine from a spot around no. 50 Boulevard du Temple – the place is marked by a plaque. As soon as the royal party were in his various lines of fire, Fieschi set the machine off, exploding bullets all over the place. He somehow missed his intended target--Louis and his children were only grazed by the hail of bullets - eighteen were killed in the mayhem and many were wounded, including Fieschi. Seemingly, he taped one gun on backwards!

Despite trying to escape, our deluded genius was captured. The king ensured that the wounded would-be assassin received the very best medical attention, so that he was fit to face justice. At his trial Fieschi decided to take advantage of this opportunity to finger every last one of his accomplices, confident that he would be pardoned, since the king had done so much to save him already. He ended up being even more surprised when he was sentenced to death by guillotine on February 19, 1836. Morey and Pépin were also executed, another accomplice was sentenced to twenty years imprisonment and one was acquitted. For some reason his death mask is on display at Norwich castle.

During that year police discovered seven plots against King Louis Philippe but none as bizarre or as lethal as that masterminded by our eighth inductee, Guiseppe Fieschi. If he still had his head, we would ask him to take a bow!

16

SIEUR FREMINET
(Dates Unknown)

TOWT and I have just returned from our annual pilgrimage to Mauritius. For the younger end of the tourist market, it is a mecca for watersports and particularly scuba diving. For some unaccountable reason, people seem to want to get close to the fish and coral that surround the island. For us viewing from a glass-bottomed boat is close enough, thank you very much.

Watching the antics of the scuba divers my thoughts soon turned to musing about who it was who pioneered the development of the support system needed to promote scuba diving and this is where our latest, somewhat shadowy, inductee to our Hall of Fame, Sieur Freminet comes in.

For those of you who are aficionados of the art of scuba diving there are two basic configurations. One is called the open-circuit where the diver passes the exhaled air into the environment and it requires them to take their breaths of air from a diving regulator. The other form is known as the rebreather where the equipment the diver uses recycles the exhaled gas, removes the carbon dioxide and compensates for the used oxygen before the diver is resupplied with gas from the breathing circuit. The amount of gas lost per circuit depends upon the design of the rebreather and the depth change during the breathing circuit.

I suppose those who wish to spend a prolonged amount of time underwater are trying to rediscover man's piscine origin and the problem of how to breathe effectively underwater has taxed many a mind, some great, some less so, over the centuries. By 1771 a British engineer, John Smeaton, had invented an air pump which when connected to a diving barrel by means of a hose allowed air to be pumped to a diver.

But 1772 saw a major breakthrough thanks to the endeavours of our French hero. Freminet developed what he called his *machine hydrostatergatique* which was a form of breathing machine which would be classified as a rebreather today. The diver was equipped with a helmet, two tubes – one for inhalation and the other for exhalation – a suit and a reservoir which was dragged by and behind the diver. In a later adaptation, Freminet strapped the reservoir to his back.

Although precise details are somewhat sketchy, it seems that our hero used his device successfully in the harbours of Le Havre and Brest for at least 10 years, judging by the inscription to a painting dating to 1784. Freminet sent six copies of the treatise he wrote publicising his machine and spoke on the subject on 5th April 1784, as the records of the Chamber of Commerce of Guienne show, noting the purpose of the invention as being in case of shipwreck or to investigate blockages in water channels.

There was, however, one inherent design flaw in the equipment. The system of inhalation and exhalation into the one tank meant that in the event of prolonged use, the levels of oxygen were depleted to dangerous levels. Alas, for our hero but the stroke of fortune that guarantees his place in our hallowed Hall of Fame, he died from lack of oxygen after being in his own device for twenty minutes.

A worthy inductee, indeed.

17

LUIGI GALVANI
(1737 – 1798)

The next inductee into our Hall of Fame is the Italian anatomy professor, Luigi Galvani. His claim to fame is his experimentation into the world of bioelectricity. In 1776 he was appointed to the Academy of Science at the University of Bologna and his duties included providing practical studies into anatomy which involved human dissections. He was also required to deliver one research paper every year, a task he fulfilled dutifully until his death.

In 1780 he was skinning a frog on a table upon which he had earlier been conducting experiments into static electricity. When his assistant touched an exposed sciatic nerve of the frog with a metal scalpel, they observed the dead frog's leg kick into life. This observation convinced Galvani of the relationship between electricity and animation and led him to develop the theory that the cause of muscle motion was electrical energy carried by a liquid and not air or fluid as had previously been thought. Our Italian scientist has been credited with the discovery of bioelectricity, a branch of science which became known as Galvanism but is now more prosaically termed electrophysiology.

Of course for every scientist who thinks outside the box, there are legions of naysayers and Galdani's theories excited the curiosity of one of his rivals, Alessandro Volta from the University of Pavia. Volta repeated our man's experiments and became convinced that the movements were due to the application of charged metal to the animal's limbs, not because of any intrinsic electric current contained within the body. Volta's experiments led him to develop an early prototype of what became the battery.

Anyway, back to our hero. The ability to bring dead limbs back to life, even if you didn't fully understand why, offered the opportunity for a bit of sport – bear in mind we are talking about a period of time when you had to take your entertainment where you found it. The demonstrations became crowd-pullers and offered the opportunity to make some cash. No one grasped the mantle more readily than Galvani's nephew, Giovanni Aldini, to whom our man entrusted the task of defending his theories. Aldini toured around Europe demonstrating the powers of Galvanism and his party piece was to carry out the experiments on human bodies.

The most famous exhibition occurred on January 17th 1803 at the Royal College of Surgeons in London when the body of an executed murderer, George Forster, was wired up to a 120 volt battery. When wires were placed on the mouth and ear of the corpse, the jaw muscles quivered and the murderer's face was seen to convey a grimace of pain. The left eye opened. The finale was to place a wire to the ear and another up the unfortunate corpse's rectum. The result was astonishing – the corpse gave the appearance of dancing, leading the Times to observe, *"It appeared to the uninformed part of the bystanders as if the wretched man was on the eve of being restored to life"*.

Galvani's theory and Aldini's experiments prompted others to try to electrify bodies in the hope that they would return to life – all experiments proving (unsurprisingly) unsuccessful – and it is thought that the craze for electric animation gave Mary Shelley inspiration for her book, Frankenstein, published in 1816.

Bioelectricity, electrical animation, the battery and Frankenstein – the legacy of Galvani makes him a truly worthy inductee to our Hall of Fame.

18

GEORGE GARRETT
(1852 – 1902)

Resurgam is Latin for I will rise again and in many ways it is an appropriate name for a submarine. However, in the era of submarine development it seems a hopelessly optimistic sobriquet.

Our latest inductee, George Garrett, was born on 4th July 1852 and was brought up in Moss Side in Manchester, the son of a curate. Having studied at Trinity College Dublin he passed the Cambridge Theological Examination and became a curate in his father's parish.

Life as a curate in those days was pretty undemanding and so Garrett had plenty of time to indulge his passion for mechanical engineering. In 1877 he had invented a diving suit which he demonstrated to the French government in the River Seine. We write elsewhere about the American experiments in submarine technology, particularly the ill-fated endeavours of Horace Hunley. Despite the set-backs, not unnaturally, interest was piqued this side of the pond and particularly in Garrett.

He was so energised to try and develop a submarine that could be used for military purposes that he formed the Garrett Submarine Navigation and Pneumatophore Company Limited – a pneumatophore was a device for removing carbon dioxide from the air – and managed to raise £10,000 from local Manchester businessmen to fund his research. He came up in 1878 with the optimistically named Resurgam which was a 14 foot hand-cranked submarine, weighing about 4.5 tons. It was nick-named the curate's egg due to its distinctive shape but its size and the fact that it could only hold one crew member meant that it was ineffective as a weapon.

Undaunted, in 1879, Garrett developed Resurgam II which was an altogether bigger affair. It was constructed of iron plates fastened to an iron frame, the centre of the vessel being made of wood secured by iron straps. It was 45 feet in length and 10 feet wide and could carry a crew of three. It was powered by a closed cycle steam engine which provided enough steam to power it for four hours. The furnace and chimney were shut off before diving – they thought of everything!

After successful trials in the Birkenhead area it was to be sailed to Portsmouth to be demonstrated to the

English navy. Unfortunately, Resurgam II was beset by mechanical problems, the crew transferred to a nearby vessel and because the hatch couldn't be secured from the outside, the vessel shipped water and sank on 25th February 1880.

You would have thought that was that but a Swedish industrialist, Thorsten Nordenfelt, was sufficiently impressed by the potential to further finance Garrett's endeavours. Together they built a submarine for the Greeks and two for the Ottomans, all of them suffering from severe stability problems. The Russians ordered a sub but it ran aground off Jutland on the way and they refused to pay.

This was the end of the partnership and Garrett emigrated to Florida where he lost all his savings in a failed farm in Florida. He then joined the US Army Corps of Engineers, rising (unlike his subs) to corporal, becoming a US citizen but dying in penury in New York in 1902.

As usual for inductees into our Hall of Fame, Garrett had an unlucky streak running though his life. However, the distinctive shape of the Resurgam is one which pretty much has been adopted by modern submarines.

George, take your place in our Hall of Fame!

19

JAMES GOODFELLOW
(1937 – Present)

There is a bewildering array of ways of paying for things these days but some of us still get an enormous amount of satisfaction in opening up our wallets and getting out some folding stuff to complete our transaction. And we don't have to queue up at a bank to get our money. All we need to do is get out a debit or credit card, insert into one of those clever machines in a wall, punch in a few numbers, making sure, of course, that no one is lurking over our shoulder attempting to memorise our Personal Identification Number (PIN) and, hey presto, not only is our card returned but some money appears, often even the right amount in vaguely convenient denominations of notes.

So embedded into our daily life is the ATM that we barely ever give it a second thought, other than when it is out of service or chews up our card, or even consider who and how it might have been developed. This is where the latest inductee into our illustrious Hall of Fame, the Scotsman James Goodfellow, comes in.

In the mid-1960s Goodfellow was working for the Glasgow firm, Kelvin Hughes, on devising a system whereby customers could withdraw cash from banks after they had closed on Saturday lunchtimes. His solution, as described in the patent application, was a system incorporating a plastic card with holes punched in it, a card reader and a series of buttons mounted into the external wall of a bank. To access their cash, the customer would insert their personalised card into the card reader of the system and punch in their PIN number – Goodfellow invented this – via a series of 10 push buttons. In other words, other than the card with punched holes, pretty much what we know as an ATM.

After Goodfellow had prototyped and demonstrated successfully his machine, he patented it on 2nd May 1966 and it was installed in Westminster bank branches in 1967. But this was not the road to fame and fortune for Goodfellow. Part of the problem was that he had a rival, John Shepherd-Barron, who worked for De La Rue and who developed a cash dispensing machine using cheques impregnated with carbon-14 which upon matching the cheque against a PIN paid out. This machine was installed at the Enfield branch of Barclays on 27th June 1967, the first publicly installed cash dispensing machine.

Shepherd-Barron was feted as the inventor of the automatic cash dispenser, something which stuck in Goodfellow's craw, and with good reason. After all, he had filed his patent some 14 months earlier. But for some reason Goodfellow kept schtum until 2005. In 2006 he was awarded an OBE for inventing the PIN. Slowly he has begun to receive the recognition due to him. The ever popular website ATMInventor.com attributes to him the invention of the ATM as we know it, although the honour for inventing the concept goes to Luther George Simjian. The Home Office now include him in the section of great British inventions in the booklet given to people aiming to secure a pot of gold aka British citizenship.

And what did he earn from his invention? A paltry $15, the standard patent signature fee being $1 a country and the patent was registered in 15 countries. Oh, and he was made redundant by Kelvin Hughes, refusing to move down south when the project was relocated.

For inventing the ATM and PIN and not receiving your just rewards, James Goodfellow, you are a worthy inductee into our Hall of Fame.

20

CHARLES GOODYEAR
(1800 – 1860)

Most people have heard of Goodyear Tyres and would naturally assume that if there was an eponymous owner of the company, he or she would have made a mint. But not a bit of it went to the inventor, which is why Charles Goodyear is a worthy inductee into our Hall of Fame.

Although there is evidence that the Mesoamericans had mastered the art of turning rubber into balls and other objects as far back as 1600 BCE, they never seemed to have seen the need for a wheel, never mind one coated with rubber. Consequently, the art of stabilising rubber was "lost" to what we call Western civilization.

Charles was interested in the possibilities presented by rubber and hooked up with a Nathaniel Hayward who ran a rubber factory in Woburn, Massachusetts. For a period of 5 years or so from 1838 until 1844, Goodyear carried out a series of experiments around Hayward's practice of drying rubber using sulphur. He made the key discovery that by heating rubber with sulphur you created a more durable and useable substance in a process which we now know as vulcanisation, named after the Roman god of fire.

How he came to make his discovery is shrouded in mystery. Goodyear, in his autobiography Gum Elastica, admitted that it was not the result of systematic scientific experimentation but rather through application and observation. It has been claimed that the discovery was the consequence of Goodyear spilling some rubber onto a stove and observing the effects of vulcanisation.

However the discovery came about, Goodyear set about raising capital to industrialise his vulcanised rubber. Unfortunately, he did not find a queue of eager investors – Goodyear had a track record of financial failure – but eventually secured the backing of the New York brothers, William and Emery Ryder. The bad luck we associate with our inductees followed Goodyear around and soon the Ryder Brothers' business failed, leaving our hero back where he started.

Some years earlier, Goodyear had set up a rubber factory at Springfield to which he moved his main business in 1842. His wealthy brother-in-law, Mr DeForest, supplied the capital to allow Charles to bounce back and he applied for a patent (number 3633) in 1844.

But misfortune continued to dog our hero. A Brit,

Thomas Hancock, an employee of Charles MacKintosh & Company, also claimed to have discovered vulcanisation and received a British patent, initiated in 1843 and finalised in 1844, weeks before Goodyear's was awarded. The dispute over primacy went to the courts and after three separate hearings, Goodyear who, had he prevailed, would have had the British patent to go with his American one and the keys to a fortune, lost.

Goodyear's death, as you would expect, was tragic. He travelled to New York City to see his dying daughter. On arrival he was told that she had already died, he collapsed and died himself shortly afterwards on 1st July 1860 at a Fifth Avenue hotel to which he had been carried.

The Goodyear Tyre Company was not founded until 1898 by Frank Sieberling. It had no connection with our hero other than using his name to honour the modern founder of vulcanisation (possibly).

Charles Goodyear, for discovering and industrialising the vulcanisation of rubber, you are a worthy inductee to our Hall of Fame.

21

WILSON GREATBATCH
(1919 – 2011)

The next inductee into our Hall of Fame is a serial inventor whose greatest contribution to the well-being of mankind was discovered by accident.

In the 1950s after leaving the navy, Greatbatch was working as a medical researcher at the University of Buffalo, trying to develop an oscillator to record heart sounds. (Un)fortunately, one day as he was assembling his box of tricks he reached out for a resistor, got hold of the wrong type and unwittingly fitted it to his oscillator. Naturally, the machine did not behave as he anticipated – instead of recording the rhythm of a beating heart he noticed that his machine gave off a rhythmic electrical pulse. The result reminded him of some discussions he had held with colleagues some time ago in which they speculated whether an electrical stimulation could make up for a breakdown of the heart's natural beats.

Showing the ingenuity that we come to expect of our inductees, Wilson decided to experiment further and two years later he had developed and been awarded a patent for the first implantable pacemaker, just two cubic inches in size.

Before Greatbatch's breakthrough, pacemakers existed but they were the size of televisions and the patient had to be wired up to them. An unfortunate by-product of the early pacemakers was that the patient often received electric shocks.

His first pacemaker was fitted into a 77-year-old who survived for 18 months. Now more than half a million of the devices are fitted per annum and the National Society of Professional Engineers recognised Greatbatch's invention as one of the top ten engineering achievements in the last fifty years.

But Greatbatch didn't finish there. He was frustrated by the limitations that battery technology imposed on his pacemaker and so in the 1970s started to develop and manufacture long-life lithium batteries. His company, Greatbatch Inc, eventually supplied 90% of the world's pacemaker batteries.

By the time of his death (I can't confirm the cause but I suspect it wasn't a heart attack) Greatbatch held over 325 patents and had made significant contributions into environmental research and AIDS.

An act of stupidity or carelessness (take your pick) ultimately made a significant contribution to medical science. Wilson, you are worthy of our Hall of Fame.

22

WILLIAM HARVEY
(1578 – 1657)

Without question medical science and our understanding of how our bodies work has developed by leaps and bounds over the last few decades. In many ways this is even more astonishing because for a millennium and a half or more, medical knowledge was in a state of stasis. Take for example our understanding of blood and how it circulates around the body.

Prior to William Harvey, the latest inductee to our Hall of Fame, our understanding of blood had been dominated by the theories of Galen, the 2nd century CE Greek speaking Roman physician. Galen believed that the veins and arteries were two separate and distinct circulatory systems, only coming into contact with each other through unseen pores. This view, at least in Western medical understanding – there were advances on this theory amongst Arab medicos but, of course, their views didn't count – prevailed until the 17th century. There was what was termed the natural system containing venous blood which originated from the liver and the vital system containing arterial blood which originated from the heart and which contained the spirits that controlled the body as well as distributing heat and life to all parts of the human body. The role of the lungs was to fan and cool the vital blood.

This view of the role of blood was blown apart by our hero. He published, in Frankfurt in 1628, a 72 page book (Exercitatio Anatomica de Motu Cordis et Sanguinis Animalibus) in which he gave a clear account of the role of the heart and the consequent movement of blood around the body in a circuit. In his opening chapter he stated that to understand the workings of the heart you really needed to observe it in action. Given the ability to observe the inner workings of the body at the time, this was pretty much impossible and Harvey despairs, saying *"...I was almost tempted to think….that the movement of the heart was only to be comprehended by God"*. Nevertheless, by theorising and the dissection of animals Harvey was able to come up with a compelling and pretty accurate explanation of how blood travelled around the body – a landmark moment in the development in the understanding of human anatomy.

Harvey was physician to Charles I and during the civil war witnessed the battle of Edgehill. His house was also ransacked by the Parliamentarian mobs and many of his papers were destroyed or scattered to the winds. He was also the first to suggest that humans and other mammals reproduced by way of the fertilisation of eggs by sperm, a theory which was not corroborated by evidence for a further couple of centuries.

Whilst all this is very worthy, what earns our hero his place in our Hall of Fame was his zeal in finding human cadavers to dissect to enhance his understanding of the human anatomy. Dissecting humans was infra-dig other than where the cadavers belonged to criminals executed by due judicial process. Harvey overcame this shortage of human material by dissecting his relatives, principally his father and his sister. You can imagine him anxiously enquiring of their health around the breakfast table and being disappointed when they own up to being in rude health, rather like, one imagines, Prince Charles and his mother. Eventually, our hero got his hands on them and medical knowledge advanced as a result.

For this cold-blooded zeal alone, William, you are a worthy inductee.

23

ELIAS HOWE JR
(1819 – 1867)

It is a device which is used everywhere many times a day around the world and is still one of the most effective means of getting a complete join between two edges of clothing. Of course, I'm talking about the zip but do you know who first had the idea that led to the creation of this everyday object? Step forward, Elias Howe Jr, the latest inductee into our illustrious Hall of Fame.

Born in 1819 in Spencer, Massachusetts, Howe is better known for his work on developing and refining the sewing machine for which on 10th September 1846 he was awarded an American patent. His version of the sewing machine contained three features which are still in use today – a needle with the eye at the point, a shuttle operating beneath the cloth to form the lock stitch – the most common form of stitch made by a mechanical sewing machine – and an automatic feed.

Of course, gaining a patent and make a commercial success of your invention are two different things and Howe struggled to get the necessary backing in the States for his machine. Our hero enlisted the help of his brother Amara who travelled over to England and was able to sell the first machine to a manufacturer of corsets, umbrellas and valises in London's Cheapside, one William Thomas, for the princely sum of £250. Suitably encouraged by this success, Elias moved his family over to England in 1848. But this ended unhappily as his wife suffered from ill-health and moved back to the States, dying the following year and Thomas proved a prickly person to deal with. Howe returned to the States almost penniless.

Although Howe struggled to make his invention a commercial reality, others had spotted the potential. Isaac Singer with the help of Walter Hunt had created a perfect copy of his machine and was selling it with the same lock stitch that Howe had developed and patented. Between 1849 and 1854 Howe was in the courts defending his patent and won a considerable amount in the way of royalties from Singer and others from his invention. Our hero spent most of his fortune buying equipment for the 17th Connecticut Volunteer Infantry of the Union Army in the American Civil War.

However, more germane to our tale is the patent which Howe applied for and received in 1851 for what he called *'An Automatic Continuous Clothing Closure'* which performed similar to the zip that we know and love today. Consisting of individual clasps which were pulled together manually and pulled shut using a piece of string, it created a gathering effect and a more complete join that a series of buttons would. Whether Howe was distracted by the legal difficulties that he was encountering over his sewing machine or whether he saw a more immediate prospect of riches from it, is unclear but whatever the reason was, Howe chose not to develop his prototype zip into a commercial proposition.

It was not until 1913 that Gideon Sandrock perfected a device consisting of interlocking oval scoops that could be joined together tightly by means of a metal slider. Even then, this more recognisable form of the zip took time to take off – the First World War finding a use for the zip for flying suits and money belts. B.F.Goodrich christened this new form of fastener the zip in the 1920s because of the sound the slider made.

For failing to recognise the potential of your prototype zip, Elias Howe, you are a worthy inductee into our Hall of Fame.

24

HORACE LAWSON HUNLEY
(1823 – 1863)

The next inductee to our Hall of Fame is Horace Hunley who over the course of his relatively short life was a lawyer, merchant, member of the Louisiana state legislature and a marine engineer. Although born in Tennessee, Horace's parents moved soon after his birth to New Orleans which is where he grew up and practised law.

His interest in marine engineering was brought to a head by the outbreak of war – the American Civil War – when he and a couple of colleagues, James McClintock and Baxter Watson, started to prototype submarines. I don't know about you but the idea of being in a submarine fills me with horror. Just imagine how scary it must have been designing and testing early prototypes.

The three collaborators worked on and funded themselves an early submarine for the Confederates, called the Pioneer. The idea behind the sub was quite compelling – by building a vessel that could travel underwater they would be able to surprise enemy shipping as, indeed, the success of subs in the Second World War demonstrated. The Pioneer was first tested in February 1862 in the Mississippi river and then was towed up to Lake Pontchartrain for additional trials. Unfortunately (there is always a vein of misfortune running through the lives of our inductees), New Orleans had fallen into the hands of the forces of the Union before trials had been completed and the Pioneer was scuttled. In 1865 it was raised and examined by Union troops and then sold off for scrap.

A second submarine built by our hero saw the light of day but sank in Mobile Bay off the coast of Alabama.

Undaunted, Huntley this time funded himself and developed a third submarine which could reach the impressive speed of 4 knots per hour, powered by a hand-driven screw. It was 3 feet 10 inches wide, illuminated by candles and its only source of oxygen was what was in the cabin when it was launched. The contraption probably presented more danger to its crew than to enemy shipping as subsequent events soon revealed. On its first trial it sank but was raised to the surface, dusted down and was ready for its second trial.

To demonstrate his faith in the vessel, Hunley was one of the crew of eight for the next trial. Unfortunately, the sub was stranded on the bottom of the sea off Charlestown in South Carolina and all the crew, including our hero, suffocated to death. It must have been a particularly gruesome way to go.

Hunley's sacrifice did not go unrewarded. The Confederacy managed to recover the vessel (presumably they removed the remains of the erstwhile crew) and sent it into action again. This time it actually sank an enemy vessel, the USS Housatonic, in 1864, the first success by a sub against enemy shipping. But the success was bitter-sweet as the sub, named the Hunley, disappeared and never returned, taking its third crew to Davy's locker. Despite this inauspicious start, the concept of a submarine took hold and was, eventually, developed into a more reliable means of marine transport.

Horace, killed by your own invention, you are a worthy inductee to our Hall of Fame.

25

MARIE KILLICK
(1914 – 1964)

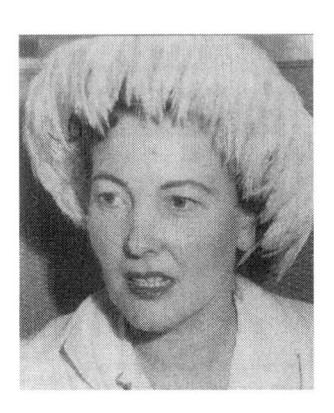

Vinyl records, we are told, are making a bit of a comeback. For the purist, the sound reproduction to be obtained from a top-end record deck beats the rather flattened and robotic sounds from a CD into a cocked hat. Of course, the purity of sound from vinyl is very dependent upon the quality of the stylus ploughing its way in ever decreasing circles through the record. And this is where the latest inductee to our illustrious Hall of Fame, Mitcham-born Marie Killick, comes in.

Having established an engineering company in Putney in 1938 which specialised in master disc cutting styli, Killick won a number of contracts after the outbreak of the Second World War to provide gem cutter disc recording equipment for use in recording air traffic and encoded transmissions. It is almost certain that she developed gem styli for playing back as well as cutting recordings during the war. In October 1945 Marie filed for a patent in respect of a stylus able to play all records and giving high quality reproduction with minimal wear.

Post war Killick's business profited from people's greater leisure time and desire to play recorded music. It was a high margin business – styli could be made for 10d and retailed for 6s. Decca made a bid of £750,000 for her business, goodwill and the British rights to her patent in 1948 but Killick rejected the overtures.

Inevitably, though, with such profits available the big boys such as Walco and Telefunken muscled in. Killick's business suffered and unaccountably she allowed her patent to lapse in 1949, just as CBS was introducing the 33 1/3 rpm record and RCA the 45 rpm single. Killick reapplied and succeeded in obtaining the reinstatement of her patent in 1951 but seemed to have been powerless or unwilling to assert her rights. That is until Pye Company of Cambridge emerged on the scene in 1953 selling a universal stylus suitable for the old 78s and the new microgroove records, boasting a stylus symmetry so close to Marie's design as to be assumed to be a copy.

Killick instructed her solicitors and so began one of the most (in)famous patent cases to be heard in the British courts. Prior to the commencement of proceedings, Pye attempted to settle for a sum so paltry that it wouldn't even cover her legal expenses. I won't bore you with the minutiae of the case – any text-book on patent law worth its salt will do a better job – but suffice to say on 20th September 1957 our inductee was awarded a certificate of validity, an injunction, costs and an inquiry for damages.

There were many curious aspects to Killick's character, not least her inattention to her business, and the prospect of a big payout – she was interviewed at the time giving a very inflated view of how much she could expect – unleashed another trait, her penchant for the high-life.

The law is fertile ground for delay and Pye, feeling hard done by by the judgement, did everything they could to delay the assessment of damages. In the meantime, Killick's riotous lifestyle meant that debts were mounting – the Sunday People ran a story in January 1959 about her scandalous lifestyle leaving poor creditors in its wake – and there was a race to see whether Marie would receive her damages before she was declared bankrupt.

The final nail was when the damages payable by Pye were assessed at £5,500 against her outstanding debts of £15,719 – her unpaid bills were famously collected up in a clothes' basket – and she had no option but to declare bankruptcy. She died in penury (just outside Guildford!) in 1964.

Next time you line a diamond stylus up against a vinyl record, give poor Marie, a worthy inductee to our Hall of Fame, some thought.

26

RON KLEIN
(C1936 – Present)

In the run up to the Festival of Mammon aka Christmas you have doubtless been giving your credit card a bit of a bash. You probably only turn it over to remind yourself of the CVV number but towards the top of the reverse of a credit card is a long metallic strip. Ever wondered who invented it?

Well step forward Ron Klein, the latest inductee of our Hall of Fame.

The Florida based inventor and Korean war veteran was working in 1964 as director of engineering for Ultronics Systems Corp. One of his clients had a bit of problem, how to determine credit cards belonging to people with a bad credit history from those with good credit. In those days, stores were given each month a list of all card holders with poor credit history. At the point of sale, the shop assistant would have to go through the ledgers to satisfy themselves that everything was hunky-dory. An inefficient and time-consuming process.

Klein exercised his grey cells on the subject and realised what was needed was some form of point-of-sale device. The machine he came up with in 1969, for which he obtained a patent, allowed the merchant to enter a credit card number into what was called a desk mountable interrogation unit, connected to a central unit memory drum. The drum would check the credit card number against its store of poor credit records and if the customer got the all clear, it would produce a carbon copy sales ticket to complete the transaction. If the check failed, a light would go off and as an extra refinement, the card would be locked from further use.

Whilst undoubtedly an improvement, it was still time consuming and Klein realised that the key to enhancing the speed of the transaction was eliminating the human intervention. In trying to make the card smarter, as we would term it today, he was attracted to the idea of utilising the Hollerith code used in computer punch cards – remember them? – which used tiny holes punched in pre-defined positions on an eighty column grid. But the process of trimming the credit card to resemble a computer input card proved expensive.

Klein then turned to the medium that was all the rage at the time, reel-to-reel tape recorders – remember them? He recognised that the magnetic tape would give the card just enough processing power to eliminate the human intervention. The quirk of the credit card's magnetic strip, its seemingly random sensitivity to the speed at which the card is swiped, is down to the tape's origin and was deployed to activate what is known as stop-start synchronisation which tells the strip to read everything after the first synchronisation and reset everything after the last impulse.

Klein, as you would expect of an inductee to our Hall of Fame, didn't patent his magnetic strip. Nonetheless it proved successful and was the basis for the enhancements which are used today.

Don't feel too sorry for Klein. Although he didn't make a mint out of the magnetic strip on credit cards, he made more than enough money from some of his other inventions. In particular, he designed a highly successful nutrition system which is used to raise chickens more efficiently.

Ron Klein, for inventing the magnetic strip on credit cards and not getting a bean for it, you are a worthy inductee to our Hall of Fame.

27

OTTO LILIENTHAL
(1848 – 1896)

The latest inductee to our Hall of Fame is the Prussian aeronautical pioneer, Otto Lilienthal.

Growing up in Anklam which is in Pomerania, Otto and his brother, Gustav, were fascinated by birds and the concept of manned flight. As boys, and rather like the Greek mythological pair, Daedalus and Icarus, they experimented with strap-on wings but, perhaps fortunately, were unable to achieve sufficient levitation to fly.

Undaunted, however, Otto, after seeing military service in the Franco-Prussian War, continued his interest in nascent aeronautics, publishing in 1889 his famous investigation into bird flight as the basis for aviation, *Der Vogelflug als Grundlage der Fliegekunst*. Otto was particularly fascinated by the flight pattern of white storks and his detailed investigations convinced him that a wing-flapping movement was the key to successful human flight.

Meanwhile, his day job was running his own company which manufactured boilers and steam engines. He devised a small engine which worked on a system of tubular boilers, much safer than many of the engines operating at the time, and this success gave him the economic freedom to carry on his aeronautical experiments.

During the early 1890s our hero cracked on at pace developing monoplanes, wing-flapping planes and biplanes. Essentially his aircraft were gliders, carefully designed to allow him to change the centre of gravity by shifting his body. However, if they caught an unexpected current of air they were difficult to control, mainly because he held the glider by his shoulders which meant only his legs and lower body could move. Throughout his career he applied for and received around 25 patents.

To begin with Otto used to jump off a small hill near Steglitz in Berlin, building a hut in the shape of a tower on the top to create a 109 metre high jumping-off point and, incidentally, somewhere to store his apparatus. In 1894 he decided to build his own artificial hill called Fliegerberg which was 15 metres tall and allowed him to jump irrespective of the wind direction.

Otto's fame spread. There were regularly crowds who assembled to watch his experiments at Lichterfelde and, internationally, he became acknowledged as a father of flight. Over his relatively short flying career he made around 2,000 flights, although his first recorded jump took him a distance of just over 25 metres. In 1893 at the Rhinow Hills he was able to cover a distance of 250 metres, a record that was still standing at the time of his demise.

Inevitably, as you would expect of an inductee, his end came in a rather spectacular fashion. On 9th August 1896 he went as usual to the Rhinow Hills and on his first jump covered his usual 250 metres. However, disaster struck on his fourth glide. The machine pitched headlong towards the earth. Otto was unable to pull the machine out of its decline and, trapped in his glider, fell to the earth from a distance of around 50 feet. He was transported by horse-drawn cart to Stolln for a medical examination where he was found to have fractured his vertebrae. Passing in and out of consciousness Otto hung on for 36 hours but even the ministrations of one of the world's most eminent surgeons in Berlin couldn't save him. His last words were, "*Opfer müssen gebracht werden!*" (Sacrifices must be made!).

Otto, for your experiments into flight and your stoicism at the end, you are a worthy inductee.

28

ANTONIO MEUCCI
(1808 – 1889)

The latest inductee into our illustrious Hall of Fame is a man who has a justifiable claim to be the inventor of the telephone, Antonio Meucci. Needless to say he lost out in his battle to secure his rights to the invention and Alexander Graham Bell saw the opportunity to scoop the prize and claim the glory.

Born near Florence, Antonio developed his first communication device in 1834 which allowed the control room and the stage at the Teatro della Pergola to communicate with each other via something akin to the pipe telephones used on ships. Migrating to the States via Cuba in 1850 he set up a tallow candle factory, the first of its kind in the country. But it was his interest in what we now know as telephony that was his main focus.

By 1856 Meucci had achieved his ambition of broadcasting his voice through wires, rigging up a device which allowed him to communicate with his bed-ridden wife on the second floor of his house from his workshop in the basement. Calling his device the *telettrofono*, he described it in his notes as "*a vibrating diaphragm and an electrified magnet with a spiral wire that wraps around it. The vibrating diaphragm alters the current of the magnet. These alterations of current, transmitted to the other end of the wire, create analogous vibrations of the receiving diaphragm and reproduce the word*". Between 1856 and 1870 he developed more than thirty different types of telephone based on his prototype. In 1860 he gave a public demonstration of his invention, reported in the New York Italian-language newspaper.

But as we come to expect of our inductees, misfortune dogged him. His candle factory got into financial difficulties, Meucci was severely burned in a steamship accident and his wife, Ester, sold his machines to a second hand shop for $6. Perhaps she was fed up with him ringing her. Nonetheless, Meucci persevered and by 1871 had secured funding from some businessmen of Italian extraction to establish the Telettrofono Company.

The next step was to file for a patent but Meucci's lawyer submitted a patent caveat, a sort of provisional application valid for a year, for a Sound Telegraph on December 28th 1871 rather than an application for full patency. To compound the error, the caveat wasn't renewed on its third expiry in December 1874 – some claim that he couldn't afford the renewal fee, a claim that should be viewed with some suspicion given that he applied for and was granted patents on other inventions, not telephone related, in the period between 1872 and 1876.

Stranger still, Meucci sent a model and technical details of his telephone to the Western Union Telegraph Company, requesting a meeting with the Company's executive. This was refused. Spotting his opportunity, Alexander Graham Bell filed for a patent on his telephone in 1876, succeeded and wrote his name into the annals of history. When the Bell Telephone Company sued Meucci and others for patent infringement, our hero requested the return of his papers, only to be told that they had been destroyed. Bell was working at the Western Union at the time. Did he have a hand in their destruction?

Meucci continued to fight but died in penury. It was only in 2002 that the US Congress acknowledged his contribution to telephony. A truly worthy inductee into our Hall of Fame.

29

THOMAS MIDGLEY
(1889 – 1944)

For those of you who are unfamiliar with the genius of the American chemist, Midgley almost single-handedly contributed to the acceleration of global warming with his discoveries and also died in a vaguely amusing way.

Midgley first applied his mind to the problem of engine knock – this is where petrol "detonates" from pressure and not spark ignition. This usually happens when there is too much oxygen in the air/fuel mixture or where the ignition timing is too far advanced. He solved the problem by identifying the petroleum additive, tethraethyl lead, or leaded petrol, what was later to become the bane of the environmentalists. There was, besides accelerating global warming, a problem with using tetraethyl lead and that was that it corroded the valves and spark plugs of engines. The ever resourceful Midgley solved that problem – but not the environmental havoc his discovery was wreaking – by employing bromine and inventing a method to extract large quantities of the stuff from seawater.

In 1930 Midgley was asked to find an inexpensive non-toxic refrigerant for use in household appliances by General Motors. He discovered that dichlorodifluoromethane, or Freon for short, fitted the bill. Freon is a major member of the CFC (chlorofluorocarbon) group of organic compounds which, too, have made a major contribution to the depletion of the ozone layer. Leaded petrol and CFCs – well done, Tom!

Naturally, all this fooling around with lead had a deleterious effect on Midgley's health. He eventually contracted polio in 1940 and lead poisoning which left him disabled and bed-ridden. Despite his disability, he remained active amongst the chemistry community and even served as president of the American Chemical Society (ACS) in 1944.

There was a softer side to the environment-wrecking chemist – he was a poet – and he concluded his presidential address to the ACS in 1944 with the following verses which seem to foreshadow his own demise, *"When I feel old age approaching, and it isn't any sport, and my nerves are growing rotten, and my breath is growing short, and my eyes are growing dimmer, and my hair is turning white, and I lack the old ambitions when I wander out at night, though many men my senior may remain when I'm gone, I have no regrets to offer just because I'm passing on, let this epitaph be graven on my tomb in simple style, this one did a lot of living in a mighty little while."*

Midgley contributed to his own demise by applying his brilliant but dangerous mind to his own predicament. He wouldn't let being bed-ridden defeat him so rigged up an elaborate system of pulleys and ropes to lift him out of bed. Unfortunately, at the age of 55 and just one month after reciting his prescient verses, he managed to asphyxiate himself after being strangled by one of the pulleys.

So, all three of his inventions contributed to his death – truly, a worthy member of our illustrious pantheon!

30

TOM OGLE
(1953 – 1981)

The road to becoming a successful inventor is paved with difficulties and our illustrious Hall of Fame is designed to celebrate those who fell by the wayside. Some were killed by their own inventions, as we have seen, some didn't either through omission, fraud or philanthropic gesture receive the rewards their inventions warranted and others either took their invention to the grave or came to a mysterious end. The latest inductee, Tom Ogle from El Paso, comes into the latter category.

With the cost of petrol consuming a significant percentage of the weekly household budget and the constant demands that we take more care over our planet by reducing our carbon footprint, a highly fuel efficient engine would seem to be just the ticket and this is what Ogle invented.

Ogle was messing around with a lawnmower and accidentally knocked a hole in its fuel tank. He put a vacuum line straight from the fuel tank to the carburettor inlet and was astonished to notice that the mower kept running and running with no discernible reduction in the amount of fuel in the tank, running an amazing 96 hours before the fuel ran out.

Our Tom then decided to see whether he could apply this technology to a car. Initial tests were disappointing until he discovered the root of the problem – when he was sucking vapour out, the fuel tank was freezing. Warming the fuel tank with heater coils, he found that he could get over 100 miles a gallon from his vehicle. The car was extensively tested, no evidence of any skulduggery was found and Ogle was courted by oil companies and financiers. Shell offered him $25 million but all of them wanted to put Ogle in a laboratory and assume the controlling interests in his patent.

Eventually Ogle signed a deal with an international financier named C.F.Ramsey which allowed him to continue to work on his device with the financier taking over the patent, distribution and development rights of the Oglemobile. In June 1978 Ramsey sold out to Advance Fuel Systems Inc but, initially, pretty much everything went on as usual. Ogle was receiving $5,000 a month plus funds for research and development and would also receive 6% of any royalties as and when they rolled in. In April 1979

Ogle opened the first of a thousand nation-wide diagnostic care centres.

But then the wheels came off. The car centre closed and Ogle's monthly payments stopped abruptly. Advance Fuel Systems informed him that he wouldn't get any royalties because they had worked on a device that produced similar results but which wasn't his. Personal tragedy soon followed, Ogle's wife and child leaving him in early 1981.

On April 14th our unfortunate inventor was shot on the street by someone who got away but survived the encounter. On August 18th he left the Smuggler's Inn drunk, staggered to a friend's apartment and collapsed. He was declared dead at El Paso's Eastwood hospital. After a cursory investigation, his death which involved a combination of prescribed pain killer, Darvon, and alcohol, was ruled to be accidental but several of those close to him indicated that they did not feel that Ogle would or could have killed himself. For conspiracy theorists, the temptation is to think that the vested interest of the oil companies had a hand in his demise.

Whether there is anything in that, when you are next filling up your gas-guzzler, spare a thought for Tom Ogle and what might have been. Tom Ogle, you are a worthy inductee into our Hall of Fame.

31

DENIS PAPIN
(1647 – 1712)

The latest inductee into our illustrious Hall of Fame, Denis Papin, is proof positive that being a genius isn't necessarily a key to fame and fortune. Sometimes life deals you a hand of cards which dooms you from the start.

Papin was born in France in Chitenay a protestant and, more importantly and disastrously, a Huguenot. His childhood was fairly conventional and from around 1661 he studied medicine at the University of Angers, graduating with a degree in medicine in 1669. Whilst working with Christiaan Huygens and Gottfried Liebniz in Paris four years later, he became interested in the idea of using a vacuum to generate motive power.

Working with Robert Boyle between 1676 and 1679 in London he developed a steam digester which was an early form of pressure cooker with a safety valve, giving a talk to the Royal Society about his invention in 1679. During the 1680s it was dangerous for a Huguenot to remain in France and so he joined his fellow religionists in Germany. The threat of persecution, if anything, heightened his creative talents and in 1689 he came up with another brain wave – using a force pump or a pair of bellows to maintain the pressure of and supply of fresh air in a diving bell. John Smeaton went on to incorporate this design in his diving bell in 1789.

Our hero's observations on the mechanical power of atmospheric pressure on his digester led him in 1690 to build the first ever model of a piston steam engine. Papin continued to experiment with pressure and steam as a means of locomotion and in 1705, in conjunction with Liebniz, developed a second steam engine which employed steam pressure. Details of the invention were published in 1707.

Whilst in Kassel in 1704, Papin constructed a ship powered by his steam engine, using paddles as its method of propulsion – the first steam-powered vessel and, indeed, vehicle, ever to be produced.

Papin was on the move again, returning to London in 1707, although it seems that he left Madame Papin in Germany. A number of his ideas and papers were read to and published by the Royal Society over the next five years but poor Papin neither received credit nor, more importantly, remuneration for his brilliance and ingenuity. Some effectively stole his ideas and used them to find their fame and fortune. The most notorious example was that of Thomas Newcomen who used Papin's 1690 description of his atmospheric steam engine to develop the first practical steam engine for pumping out water. We remember Newcomen, not Papin.

The last definitive mention of our hero was in January 1712. He was on his uppers and bemoaning his fate. He died, it is thought, later that year and was buried unceremoniously in an unmarked pauper's pit.

Denis Papin, as a pioneer of steam locomotion, you are a worthy inductee to our Hall of Fame.

32

ARTHUR PAUL PEDRICK
(1918 - 1976)

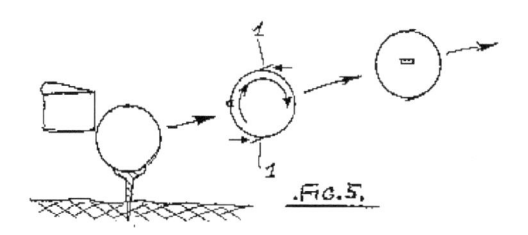

The latest inductee into our Hall of Fame is Arthur Paul Pedrick who between 1962 and his death in 1976 filed 162 United Kingdom patents. Some people are cursed with an overactive imagination and come up with ideas which, as in Pedrick's case, are centuries ahead of their time.

Little is known about our hero but he seems to have served during the Second World War as a temporary Engineer Lieutenant and from 1947 to 1961 worked for the Patent Office when he was sacked for alleged inefficiency. It was then he set about his career as an inventor.

His inventions were often if not always regarded as unworkable and none are known to have seen the light of day. Those of a charitable disposition would say he was ahead of his time, particularly in his concern to solve some of the major environmental problems. Those of an uncharitable disposition might conclude he was seriously unhinged. I will list a few of his inventions with their Patent references and leave it to you, dear reader, to decide.

For those of you who while away the hours on a golf course, one of the major problems you will encounter is controlling the spin of the golf ball. Step forward Pedrick with the answer to your problems. The ball (GB1121630) has flaps cut into its exterior which are normally held flush by magnets. If the ball has been mishit and is spinning, the centrifugal effect overcomes the magnetic force and the flaps project to overcome the spin. An additional feature is that the ball reflects radio waves to a homing device carried by the player making the ball easier to find if it disappears into the rough. Simple. Why did it not catch on?

Next up is his solution to solve the irrigation problems in the Australian outback (GB1047735). This is pure genius and involves pumping snow and ice balls from Antarctica. The snowballs accelerate under gravity from the Antarctic plateau some 10,000 feet high, reaching a speed of some 500 miles per hour when they get to sea level. Using Coriolis force (no I don't know what it is either) the balls are piped naturally using the Earth's gravitation to the outback where they can be used as water.

For those of us who work in tall buildings, a nightmare is the building catching fire and we cannot escape. Naturally, Pedrick had a solution for this (GB1453920). Rolled up fire curtains are fitted at roof-level. When the building catches fire, the curtains are released to envelope the sides of the building and extinguish the fire. The curtains were fitted with apertures allowing some air in to enable the occupants of the building to breathe.

And finally, his horse powered car (GB1405755). A horse was strapped to the back of the car with a feed box. To control the speed at which you travelled, the car's accelerator pad varied the thrust that the poor horse had to exert to reach the feed box. To brake, the brake pedal was linked to the horse's halter. The ignition switch gave the horse a mild electric shock to stimulate it into movement.

Whether mad or a genius, Pedrick is a truly worthy inductee into our Hall of Fame.

33

ANTHONY E. PRATT
(1903 – 1993)

The latest inductee to our Hall of Fame is the wonderfully named Anthony E. Pratt whose surname, as events revealed, was quite appropriate.

Born in August 1903 in Balsall Heath, a suburb in Birmingham, our hero left school at the age of 15 to pursue a career in chemistry. However, he was also a very accomplished pianist and with the absence of formal chemistry qualifications blocking his progress in his chosen career, he decided to concentrate on honing his musical skills and earned a living playing piano recitals in country hotels and on cruise ships.

During the Second World War Pratt worked in an engineering factory, the dull (but vital) work giving him time to think.

The inter-war years were the golden age of the detective novel – Agatha Christie, Raymond Chandler, Dorothy L Sayers etc. – and quite often the soirees that Pratt played at also staged murder mystery extravaganza. With no television and restrictions on movement in force, evenings during the War were deadly dull. Pratt thought that a way to enliven evenings would be to develop a board game involving the unmasking of a murderer. A country house with all its sprawling rooms – so often the setting for a whodunit – could be hosting some form of evening entertainment. One of the guests could be found murdered, all the guests could fall under suspicion and by putting clues together the players of the board game could work out who the murderer was and how the crime had been committed.

And so the seeds of a game germinated in Pratt's brain and during the course of 1943 he and his wife, Elva, had designed a game which was sufficiently well-developed that by 1st December 1944 he had filed for a patent. The game was originally called Murder and Elva designed the artwork that went along with the game.

The game was taken up by Waddingtons, who after a few minor modifications and a rebranding of the game to Cluedo – and amalgam of the word "Clue" and the Latin verb "Ludo", to play – launched it in 1949. Sales weren't brisk and by 1953 Waddington offered Pratt the princely sum of £5,000 for all the overseas rights of the game. Unfortunately, Pratt took the silver, the equivalent of around £106,000 in today's money, but in the process he denied himself access to the millions that would have rolled as overseas sales picked up.

The British patent expired in the early 1980s and so the money from what became a pre-eminent parlour game dried up. Pratt died at the age of 90 in 1993 having suffered in his later years from Alzheimer's disease.

Anthony, for developing a favourite game of mine, Cluedo, and foolishly tossing away the key to untold riches, you are a worthy inductee to our Hall of Fame.

34

LOUIS LE PRINCE
(1841 – 1890)

The latest inductee into our illustrious Hall of Fame is Metz born Louis le Prince, the forgotten founding father of what is now the multi-billion-dollar film business.

Our hero in his youth spent time in the studio of his father's friend, Louis Daguerre, from whom he received lessons in rudimentary photography and chemistry. He went on to take a post-graduate degree in chemistry at Leipzig University. Moving to Leeds in 1866 to join a firm of brass founders called

Whiteley's and marrying his boss' daughter, Elizabeth, le Prince and his wife earned some renown for fixing photographs on to metal and pottery and some examples of their work, including portraits of Queen Victoria and William Gladstone were sealed in a time capsule and placed in the foundations of Cleopatra's Needle on the banks of the Thames.

Moving to the United States, le Prince pursued his interest in developing a camera capable of shooting moving pictures. In 1888 he was granted a dual-patent on a device which contained 16 lenses and was capable of taking moving pictures and projecting the results. It was not altogether successful, though, because each lens captured the image from a slightly different angle and the result was rather jumpy.

Undeterred, le Prince returned to Leeds and built a single lens camera which, on 14th October 1888, he used to film the world's first motion picture, now known as the Roundhay Garden Scene and then went on to record scenes of trams, horse-drawn vehicles and pedestrians crossing Leeds bridge. A blue plaque today marks the spot where he had set up his camera. The films were soon shown at what would be the world's first picture show, in Leeds.

But, as is the way with our inductees, le Prince was never able to enjoy his success. In September 1890 he was due to travel to the States to give a public demonstration of his invention and decided, whilst in France, to visit his brother in Dijon. He boarded the train but was not there when it arrived in Paris, nor was his luggage. He was never seen again and was officially declared dead in 1897. A search of Parisian police archives in 2003 revealed a picture of a drowned man dating to 1890 which looked uncannily like le Prince.

Inevitably, there are a number of theories as to what happened to him. One theory, espoused by his grandson, was that the business was in financial difficulties but, au contraire, it seemed that it was profitable and le Prince had high hopes for his camera. Another is that he eloped because he was gay and feared being outed. Another theory is that he was murdered by his brother, who was the last person to see him alive.

The most intriguing theory, though, involves the Steve Jobs of his time, Thomas Edison. Perhaps le Prince was assassinated? Not long after le Prince's disappearance, Edison tried to assert his claim that he had invented cinematography and thus was entitled to reap the rewards of the invention. This claim was disputed in the American courts by the American Mutoscopy Company. Le Prince's son, Adolphe, was due to be called as a witness to demonstrate his father's two cameras but was never called and the case was decided in Edison's favour, a decision which was overturned a year later.

But the damage had been done. Edison was able to pass himself off as the inventor of cinematography and le Prince's contribution was not officially recognised until 1930. Adolphe, too, died in mysterious circumstances in 1892, having been found dead on a duck shoot.

Louis, for your contribution to cinematography, you are a worthy inductee.

35

ROBERT RECORDE
(1512 – 1558)

The equals sign (=) and the plus sign (+) are such fixtures in our system of mathematical notation that it is sometimes difficult to conceive of how we coped without them. Some clever bastard must have devised them and this is where the latest inductee into our Hall of Fame, Robert Recorde, comes in.

Our Robert was born into a respectable family in Tenby in Pembrokeshire in 1512 and entered Oxford University in 1525. He was elected a fellow of All Souls College in 1531 and decided to specialise in what passed for medicine in those days. He transferred his allegiance to Cambridge University where he studied for a M.D in 1545. Returning to Oxford, he taught medicine and then moving on to London became the physician to King Edward VI – even his ministrations couldn't halt the untimely demise of the sickly boy king whose passing heralded the reign of Bloody Mary – and also became controller of the Royal Mint and Comptroller of Mines and Monies in Ireland.

But it was as a mathematician that Robert achieved lasting fame. Prior to his development of the now customary notation for equals, mathematicians had to write out the Latin word aequalis, meaning equals, every time. This of course became very tedious if you were doing any form of extensive calculation. In his book "The Whetstone of Witt" published in 1557, Recorde acknowledged the inadequacy of the then current form of noting equality and proposed the adoption of what he called the Gemowe lines, from the Latin gemellus meaning twin, "*to avoid the tedious repetition of these words: "is equal to", I will set (as I do often in work use) a pair of parallels, or Gemowe lines, of one length (thus =), because no two things can be more equal*".

The notation he developed was wider than that which we use today but nonetheless it was a stroke (or perhaps two) of genius. Not everyone agreed with Robert though and mathematicians, who seem to have been a conservative lot, did not universally adopt the notation until the 18th century. Some preferred to use the symbol || whilst others persisted with the abbreviations ae or oe, denoting aequalis. Eventually, the notation became the norm.

But, as we have often found with our inductees, genius does not always equate with happiness.

Recorde was sued for defamation by a political enemy in the turbulent times following Mary's accession and found himself languishing in the debtors' prison in Southwark, the King's Bench Prison, where he subsequently died.

Robert, for introducing the Gemowe lines, you are a worthy inductee to our Hall of Fame.

36

SYLVESTER H. ROPER
(1823 – 1896)

I am always fascinated by individuals who have an inventive streak in them. The latest inductee to our illustrious Hall of Fame is just such a chap.

Born in Cambridge, Massachusetts on 24th November 1823 to a cabinet-maker, the young Sylvester exhibited a mechanical bent from an early age. By the age of 12 he had made a stationary steam engine, even though he had never seen one before. When he was 14 he had built his first locomotive engine, sometime before he saw a real life example in Nassau.

Married in 1845, our hero moved to Boston in 1854. It was around this time that he invented a Hand-stitch Sewing machine and later a machine for making screws and a foldable fire escape. In 1863 Roper had built his first steam carriage, one of the very earliest automobiles and was seen driving it around the streets of Boston, no doubt to the amusement and consternation of bystanders and pedestrians. One version of his 1863 carriage found a resting place in the Henry Ford museum.

As well as four wheels, Roper was fascinated by the possibilities of harnessing steam power to the bicycle which was beginning to gain some traction as a popular form of transportation. The Roper steam velocipede which saw the light of the day shortly after the ending of the Civil War may well have been the first motorcycle. For this invention, Roper was inducted posthumously into the Motorcycle Hall of Fame in 2002.

Roper did not just concentrate his attention and talents on locomotion. He found time to invent the first shotgun choke, which was a series of tubes which could be threaded into or removed from the outside of a shotgun barrel to enable the gunman to vary the spread of the shot to suit the range and size of target that was in their sights. On 4th April 1882 he and Charles Miner Spencer applied for a patent for a repeating shotgun mechanism and in his own right three years later, Roper applied for a patent for an improved shotgun loading mechanism.

But it was locomotion and velocipedes that were his first and real love. His early prototypes – essentially a small steam engine attached to a bone shaker, requiring both coal and water – were impractical but by the 1890s he had developed a compact engine which could be attached to a safety bike. At the time there were over 500 bicycle manufacturers in the US and big money was to be had from winning bike races.

On the fateful day of 1st June 1896 Roper, aged 73, raced his steam bike against professional riders who could not keep up with him. He clocked a mile in 2 minutes 1.4 seconds – average speed 40 mph – but was seen to wobble and then fall off the track, hitting his head and dying. The autopsy showed that the cause of death was heart failure, although it could not be established whether the crash was the cause of the heart failure or whether it was the other way round.

Sylvester, for your pioneering work in locomotion, you are a worthy inductee into our Hall of Fame.

37

JEAN-FRANÇOIS PILÂTRE DE ROZIER
(1754 – 1785)

The latest inductee to our prestigious Hall of Fame is the French physicist, chemist and aeronaut, Jean-Francois de Rozier.

Rozier was born in Metz which was an important garrison town on the French border. The proximity of a large military hospital to where he grew up sparked his interest in chemistry and pharmaceuticals. Rozier taught physics and chemistry at the Academy in Reims where he came to the attention and was taken under the wing of Louis XVI's brother, the Comte de Provence.

It was while under the Comte's patronage that our hero conducted research and published some papers on optics and the cause of thunder. He worked on a state-sponsored commission whose brief was to alleviate the noxious airs that pervaded the French capital at the time and he came up with a breathing apparatus, which was a bit like scuba equipment, to assist workers in circumstances where the air was harmful to their health.

But Rozier's real claim to fame was his pioneering work in aeronautics. His imagination was fired when he witnessed the first public demonstration of a balloon by the Montgolfier brothers in June 1783. Rozier muscled into the act assisting in the first untethered balloon flight, from the lawn of the Palace of Versailles on 19th September 1783, the passengers being a sheep, a cockerel and a duck.

Consideration was given to a manned flight and the initial idea was that given the dangers of the venture, the human passengers should be expendable convicted criminals. *"Au contraire,"* cried our hero, the passengers should be from the nobility and he persuaded the king that he and the Marquis d'Arlandes be allowed the (dubious) honour of being the first passengers.

Their first tethered ascent was on 15th October 1783 and on 21st November they took off on the first untethered flight, lasting 25 minutes and attaining a height of 3,000 feet. This made Rozier an international sensation, lauded by poets and heralded as a hero throughout Europe. But his star soon waned as others rose to the challenge.

On 26th July 1784 Jean-Pierre Blanchard (and Dr John Sheldon) had stolen our hero's thunder by flying from Blighty to France. Undeterred, Rozier decided to attempt a crossing from la belle France to Blighty, a much trickier journey because of the adverse winds. For the attempt Rozier designed a special balloon, which was a combination of a hydrogen and hot air balloon to give it more lift.

On 15th June 1785 Rozier and his companion, Romain, set off from Boulogne-sur-Mer. After travelling 5km the wind suddenly changed, causing the balloon to suddenly deflate and catch fire, plunging out of control 1,500 feet up in the air. *Eheu*, both occupants were killed, becoming the first fatal casualties in aeronautical history. For his bravery Rozier's family was awarded a pension by the king.

The modern hybrid gas and hot air balloon is called the Roziere balloon in his honour.

Rozier, for your pioneering spirit and for your Icarian downfall, you are a worthy inductee.

38

JONAS SALK
(1914 – 1995)

Whilst some cannot resist the lure of making a tidy profit selling so-called cures for every day maladies, some see the improvement of public health as a moral commitment from which it is inappropriate to profit. The latest inductee into our illustrious Hall of Fame, Jonas Salk, stood firmly in the latter camp.

Thanks in no small part to Salk's endeavours it is hard today to recall what a threat to the health of children polio was in the early to mid 20th century. It was a resilient disease, epidemics coming in waves, its victims principally children, and by 1952 it was killing more than any other form of communicable disease.

It was known that the disease was transmitted by faecal matter and secretions of the nose and throat, entering the victim orally, establishing itself in the intestines and then attacking the brain or spinal cord but there was little progress on finding a vaccine.

Salk, a New Yorker, who was working as a virologist at the University of Pittsburgh got his big break in 1948 when he was invited by Harry Weaver to join the National Foundation for Infantile Paralysis to help find an effective vaccine. Money poured in to fund the research with many of the scientists engaged in research concentrating on developing vaccines from live virus. The results were disappointing and deadly – in one experiment involving live virus, six children were killed and three left crippled.

Salk decided to take another route, concentrating on developing a vaccine using the safer killed virus. Tests were conducted on laboratory animals, successfully, and on July 2nd 1952 he inoculated 43 children at the compassionately named D.T Watson Home for Crippled Children and some weeks later more children at the Polk State School. To give added confidence to his vaccine Salk immunised himself, his wife and his three children. The trials were extended to include around one million children, known as the polio pioneers, in 1954. The results were astonishing - no one who had received Salk's vaccine became infected with polio and they all produced antibodies, those who had previously had polio generating the most.

On April 12th 1955 an announcement was made at the University of Michigan that the vaccine was safe. Such was the concern about polio that the event turned into a bit of a media scrum with 16 newsreel cameras at the back of the auditorium and Eli Lilly and Company reportedly paying $250,000 to broadcast the event. When the announcement was made, many in the auditorium broke down in tears in relief. Church bells rang and prayers were said. One shopkeeper famously painted the slogan, *"Thank you Dr Salk"* on his shop window.

Salk, a naturally retiring man, was showered with honours and awards, his vaccine being described as a victory for the nation. More importantly, Salk's vaccine was put into commercial production and within two years, by the summer of 1957, 100 million doses had been distributed across the States and other countries followed suit. Reported complications from administering the vaccine were remarkably few and far between and countries that adopted Salk's vaccine were soon reporting that polio was a thing of the past. Other countries which had eschewed the vaccine still reported cases of polio.

Salk didn't patent the vaccine - whether an application would have succeeded owing to the standards of the time is debatable - and when asked why, he responded, *"Would you patent the sun?"* For that Jonas, you are a worthy inductee into our Hall of Fame.

39

CARL WILHELM SCHEELE
(1742 – 1786)

The latest inductee to our Hall of Fame is the chemist, Carl Scheele, who came from the then Swedish occupied region of Pomerania. Such were the misfortunes that befell him during his life he has been dubbed hard-luck Scheele.

From a young age Scheele was fascinated by gases and conducted experiments into the composition of air. In those days a substance called phlogiston was supposed to be released from any burning material. Once it had been released and consumed, the combustion would stop. Air was thought to be an element in which chemical reactions took place but which did not interfere with the reactions. Our hero carried out a number of experiments in which he burned substances such as saltpetre, manganese dioxide and heavy metal nitrates such as mercuric oxide. His experiments led him to conclude that air was composed of two gases – what he termed fire air and foul air – and tried to explain the fire air which he had isolated, what we know as oxygen, in phlogiston terms.

Alas for poor Carl by the time he published his findings in 1777 he had been beaten to the draw by Joseph Priestley and Lavoisier who had both published their results into the composition of air and into the isolation of oxygen.

As well as oxygen, Scheele isolated other chemical elements such as barium and manganese (in 1774), tungsten (1781) and a number of chemical compounds including citric and lactic acids, glycerol, hydrogen cyanide and hydrogen sulphide. Our hero discovered a process which was similar to what we now know as pasteurisation and in 1769 had developed a method that enabled phosphorus to be mass-produced, thus enabling Sweden to become the leading producer of matches.

Another missed opportunity for our hero centred around his experimentation with hydrochloric acid. When treating pyrolusite with the acid over a warm sand bath, he noticed that a strongly smelling green gas was produced which was denser than ordinary air. He spotted that the gas stripped out colour and named the substance he had discovered dephlogisticated muriatic acid. Sir Humphrey Davy renamed it chlorine and took the prize for isolating it, chlorine becoming the foundation for disinfectants.

Of course, being at the cutting edge of chemical experimentation meant that you often didn't have a real clue about the properties of the elements and compounds you were isolating. In his zeal to understand and describe precisely the end products of his experiments, he would often sniff or even taste what had been produced. We now would readily recognise that this inadvertent and unnecessary exposure to noxious substances such as arsenic, mercury, lead and various acids would do you no good. And so it came to pass. At the early age of 43 he took to his bed at his home in Koping and died as a result of what the medics diagnosed as mercury poisoning, the victim of his experimentation.

Carl Scheele, for being the unsung discoverer of oxygen, manganese and chlorine (amongst other substances) and being a martyr to your thirst for knowledge, you are a worthy inductee.

40

WALTER L. SHAW
(1916 – 1996)

The latest inductee into our illustrious Hall of Fame is New Jersey born telecommunications engineer and inventor, Walter L Shaw. His tale is sad and cautionary.

Modern businesses and governments wouldn't survive without some of the telephonic devices we take for granted like speaker phones, call forwarding, conference calling, secure lines and the insurance industry would be the poorer without burglar alarms which connect directly to the local cop shop. So embedded are these devices into our daily life that we barely give them a second thought, let alone consider which genius may have invented them. This is where Shaw comes in.

He started his working career in 1935 at Bell Laboratories but during his spare time tinkered away pursuing his ideas and theories around telephony. His first invention in 1948 was an automatic loud-speaking hands-free telephone aka a speakerphone. In 1953 he invented a two-way communications unit. In 1954 he came to the attention of the American president, Eisenhower, and was commissioned to develop a secure telephone link connecting the White House to the Kremlin, known as the Red Phone.

But Shaw's genius didn't end there. He invented an automatic re-routing system in 1968, something we now know as call-forwarding, in 1969 conference-calling equipment and in 1971 a Remote Dialling Apparatus with Encoder and Decoder. He was also responsible for developing a tone generator which was later to provide us with our now so familiar touch tone dialling facility. His later inventions included voice print recognition and a burglar alarm that dialled the police. In all, he held 39 patents.

But Shaw's ingenuity didn't earn him untold riches. His problems were twofold. Firstly, AT&T held a monopoly on telecommunications in the States at the time and they were somewhat miffed that an employee had developed all this wizardry in his own time. They wanted the patents and rights to his inventions and when he refused, he left their employment. But he didn't have the resource to put any of his inventions into commercial production. This meant that when the patents expired, the coast was free for others with greater resources to exploit the fruits of his labours.

There was some interest from another source for his black box, which allowed long distance calls to be made free of charge and untraceable – the Mafia. They saw the potential for its use in bookmaking and other illegal activities. But even with the Mafia on his side, Shaw's fortunes didn't improve. Rather he came to the attention of the authorities, was hauled in front of a Senate subcommittee and in 1975 was arrested and the following year convicted on eight charges of illegal phone usage. He spent in excess of 11 years in jail.

He died from cancer in 1996. His son, Walter T Shaw, was so embittered by his father's treatment that he pursued a career of crime, becoming one of the world's most notorious jewel thieves with over 2,000 robberies to his name.

And the black box, perhaps, did find a more legitimate usage. The blue boxes sold by Steve Wozniak and Steve Jobs to allow free telephone calls to be made were so similar to Shaw's invention that many have claimed there to be a clear connection. As Jobs said somewhat coyly, *"if it hadn't been for the blue boxes, there wouldn't have been an Apple"*.

Walter, for your enormous contribution to telephony as we know it today and for your ill-treatment at the hands of corporate America, you are a worthy inductee into our Hall of Fame.

41

HENRY SMOLINSKI
(Died 1973)

We live in a world full of multi-functional devices. You are probably reading this on a device on which you can make phone calls, surf the internet, play videos, games and listen to music. The desire to kill two (or more) birds with one stone goes back a long way – just think of the Swiss Army knife.

And there are some frontiers to crack. Just take transportation. A car is great from getting from A to B and is very manoeuvrable. However, it is inefficient or useless if you have to travel great distances or there is an obstacle like a mountain range, a sea or an ocean in the way. An aeroplane conversely is great for long distances and overcoming geographical obstacles but doesn't have the manoeuvrability of a car. Why not combine the two into one?

Step forward the latest inductee into our Hall of Fame, Henry Smolinski.

Smolinski was an engineer by profession and was trained at Northrop, a major American aircraft manufacturer. He left to form a company of his own, Advanced Vehicle Engineers, which was focused on bringing a car which flew to the unsuspecting public.

In 1973 he had developed his first two prototypes, achieved by fusing the rear end of a *Cessna Skymaster* with a *Ford Pinto* car. The tail section was designed to be detachable. It was intended that both the aircraft's engine and the car's engine would be used for take-off, thus shortening the take-off roll. Once airborne, the car engine would be turned off. Once the machine had landed, the four-wheel braking system would stop it within 160 metres or less. Telescoping wing supports would be extended and the airframe tied down like any other aircraft. The Pinto could then be unbolted and driven off like any other car. Simple – what could go wrong?

The intention was to put the mad-cap design into production in 1974 – prices would range from $18,300 to $29,000 - but, not unnaturally, it needed to be tested.

On 26th August 1973 test pilot, Charles Janisse, took off from Camarillo airport in California but soon aborted and ditched the machine in a bean field, reporting that the right wing strut base mounting had failed.

Undeterred, Smolinski himself conducted the next test flight on, ominously, 11th September. Again the right-wing strut detached from the Pinto but this time with an inexperienced pilot at the helm, the wing folded as the novice pilot tried to turn the craft. The result was a catastrophic crash which resulted in the deaths of Smolinski and his associate, Harold Blake.

An investigation by the National Transportation Safety Board reported that bad welds were partly responsible for the crash as well as poor design. So the flying car is still a pipe-dream but at least Smolinski was prepared to put himself at risk to further technological advance – an attribute that makes him a worthy member of our mad-cap Hall of Fame.

42

PERCY SPENCER
(1894 – 1969)

One of the things that has revolutionised the kitchen and accelerated the acceptability of convenience foods is the microwave oven. I remember buying my first one in the 1980s and was astonished how heavy it was as I carried it from the shop to my flat. What many people don't know is who invented the microwave. This is where the latest inductee into our illustrious Hall of fame, Percy Spencer, comes in.

Born in Howland, Maine, by 1939 Spencer was one of the leading experts in radar tube design, working for Raytheon. Magnetrons were used to generate the microwave radio signals that were fundamental for radar and Spencer developed a more efficient way of manufacturing them, by punching out the parts and soldering them together rather than using machined parts. This meant that the rate of production increased from a stately 17 per day to around 2,600.

Whilst standing by an active radar set, our Percy noticed that a chocolate bar in his pocket had melted. He decided to investigate the phenomenon by experimenting to see the effect that exposing various foodstuffs to a magnetron would have. Some popcorn kernels became the world's first microwaved popcorn. More amusingly, an egg was placed in a kettle and the magnetron was placed directly above it. To the doubtless consternation of one of his colleagues who was peering over the contraption to see what was going on, the egg exploded in his face.

Undaunted by this set back Spencer persevered and before long had produced a contraption which consisted of a metal box to which a high density electromagnetic field generator was attached – the world's first microwave oven. The magnetron sent microwaves into the metal box, trapping them and enabling them to be used in a controlled and, mercifully, safe environment. His experiments demonstrated that not only could food be cooked in the microwave oven so that they were edible, it could be done much more quickly than in a conventional oven.

His employers, Raytheon, applied for a patent for his oven, the Radarange, on 8th October 1945. A prototype was installed in a restaurant in Boston and by 1947 the first commercially available microwave oven was launched on to the unsuspecting public. They were around 6 feet tall, weighed 750lbs and phenomenally expensive, retailing at around $5,000 a time. The magnetron had to be water-cooled which meant that the device had to be plumbed in.

Not unsurprisingly, initial sales were disappointing but soon after further refinements and modifications, an air-cooled, lighter oven was developed. Not only was it cheaper – retailing at around $2,000 to $3,000 but it didn't require the services of a plumber to install. The food industry began to twig on to the advantages of a microwave, allowing them to keep refrigerated food up to the point that it was required and then heat it up, resulting in fresher food, less waste and financial savings.

By 1967 the first counter-top, 100 volt domestic oven was available, costing $500. The take-up was phenomenal and by 1975 sales of microwaves had exceeded those of more conventional gas-powered ovens. And the rest is history.

As for Percy, whilst he climbed up the greasy corporate pole at Raytheon, ending up as a Senior Vice President and Board member of Raytheon, he didn't receive a share of the royalties. All he got was $2, the standard gratuity paid by Raytheon to employees who invented things.

Percy Spencer, for inventing the microwave oven and not sharing in the financial success of your product, you are a worthy inductee into our Hall of Fame.

43

HUGH EDWIN STRICKLAND (1811 – 1853)

The latest inductee to our Hall of Fame is Hugh Strickland who made his name as a geologist and ornithologist. As a boy he became interested in natural history and whilst at Oriel College, Oxford he attended the lectures of John Kidd on anatomy and William Buckland on geology. Their passion for their subjects encouraged the young Strickland to develop his passion for zoology and geology.

After graduating with a BA in 1831 and a MA the following year, our hero returned to his home near Tewkesbury and began to study the geology of the Vale of Evesham, sending papers to the Geological

Society of London in 1833 and 1834. He continued with his interest in birds and having been introduced to William Hamilton, accompanied him in 1835 on a trip through Asia Minor, the Thracian Bosphorus and to the Greek island of Zante. The trip was very productive and inspired Strickland to write and present to the Geological Society the following year (1836) papers on the geology of the areas. He also published a book in 1842 describing the results of his journey and subsequent trip to Armenia, entitled *Researches in Asia Minor, Pontus and Armenia*.

In 1842 he was commissioned by the British Association to consider developing rules for zoological nomenclature. His report went on to develop a codification based on the principle of priority which to this day is the fundamental guiding principle for biological nomenclature. In his researches Strickland did much pioneering work into the grouping and classification of birds – the result, a chart consisting of bits of paper stuck together with circles in paint identifying the groupings – and it is clear that his analytical approach took him to the cusp of realising that birds (and, by extension, all fauna) were part of an evolutionary process. He also wrote a book, with Alexander Melville, on the Dodo and other extinct birds of Mauritius and the area in 1848.

All very worthy and by 1852 he had been elected a Fellow of the Royal Society. But his claim to our Hall of Fame rests upon his untimely demise. "Poor Hugh" as he came to be known had been attending a meeting of the British Association at Hull in 1853. Having stopped off to view Flamborough Head he then went off on 14th September 1853 to examine the railway cuttings of the new Manchester, Sheffield and Lincolnshire railway near Retford.

Standing a little way up the track just beyond a tunnel to make a sketch of the strata revealed by the excavations, he stepped back from the down-line on to the up-line to let a slow coal train go past. Being so near the tunnel and the noise of the coal train being deafening, our hero was unaware that he had stepped into the path of an express train. Despite the driver's frantic attempts to stop the train, Strickland was struck and died instantaneously. His gold watch stopped upon impact, showing the time to be 20 minutes past 4. So notorious was his death that it was still being used as a warning to geology undergraduates in the 1980s not to examine railway cuttings.

Hugh Strickland, for being a martyr to your love of geology, you are a worthy inductee!

44

JOSEPH SWAN
(1828 – 1914)

The latest inductee to our illustrious Hall of Fame is a man who didn't receive the fame and riches that his inventiveness deserved. Step forward Joseph Swan.

Born near Sunderland, Swan exhibited an enquiring mind as a child and improved himself as was the fashion in those days by attending lectures at the Sunderland Athenaeum and became a partner at a firm of manufacturing chemists owned by his brother-in-law, John Mawson. By the 1850s Swan began experimenting with carbon filaments in an attempt to develop an electric light bulb. But his endeavours came to naught as the vacuum pumps available at the time were unable to remove enough air from the lamps to make them work.

Improvements in the technology of vacuum pumps encouraged Swan to have another go around 1875, using carbonised paper filaments in an evacuated glass bulb. As there was little residual oxygen in the vacuum tube to ignite the filament, it glowed with almost white-hot intensity without catching fire. The problem, though, was that the bulb needed heavy copper wires to supply the filament which had low resistance.

By late 1878 Swan was sufficiently encouraged to report his success to the Newcastle Chemical Society and on February 3rd 1879 demonstrated his working lamp to an audience of over 700 people. Swan then sought to improve the carbon filaments and the means of attaching its ends by devising a means of treating cotton to produce what was called "parchmentised thread". He applied for and secured a British patent. Swan's house, Underhill, in Kells Lane near Gateshead was the first in the world to be fitted with an electric light and the Lit & Phil library in Newcastle was the first public room to be so lit. In 1881 Swan established the Swan Electrical Company and began commercial operations.

The problem with Swan's bulb, though, was that the gasses trapped in the rod when the light was activated were released and caused a dark deposit of soot to build up on the inner surface of the bulb. This meant that they had a relatively short life and were largely impractical. Across the pond and independently of Swan, Thomas Edison, was working on his variant of a light bulb which improved the life of the bulb by deploying very thin filaments with high electrical resistance. Because of the high resistance they only needed a relatively small current to glow – his Bristol-board lamps produced in late 1879 lasted about 150 hours and the bamboo lamps of 1880 glowed for 600 hours.

Though Swan had invented his bulb some months before Edison, Edison was more aggressive in asserting rights and secured patents in America for what were pretty direct copies of Swan's bulb. Swan resisted a patent challenge in the courts from another lamp maker. Edison decided to work with Swan rather than against him, creating the Edison & Swan Electric Light company in 1883. Edison, however, retained the patent rights in the States, a much more lucrative market, and whilst the company sold bulbs with a cellulose filament that Swan had invented, in 1883 Edison's own company continued with bamboo filaments until the creation of the General Electric Company in 1892 when cellulose became the standard.

Swan was not as commercially minded or aggressive as his American rival and partner and so his achievements were over shadowed. But Swan did receive a knighthood in 1904, was awarded the Royal Society's Hughes medal and received the French Legion *d'honneur*. Small reward indeed for the true inventor of the light bulb.

For that, Joseph Swan, you are a worthy inductee into our Hall of Fame.

45

MAX VALIER
(1895 – 1930)

Our latest inductee is Max Valier who, despite a relatively short life of 35 years achieved more than enough to earn his place in our Hall of Fame.

Born in Bozen (Bolzano) in the South Tyrolean region of Austria whilst the Austro-Hungarian empire was still wheezing away, Valier enrolled into the University of Innsbruck at the age of eighteen to study physics. In order to earn some money to fund his studies he trained as a machinist in a nearby factory. Unfortunately, however, the outbreak of the First World War disrupted his studies – he never resumed them – and he fought for his Emperor by seeing service in their nascent army air corps as an aerial observer.

Upon his return to 'civvy street', Max became a freelance science writer. His Road to Damascus moment came in 1923 when he read Hermann Oberth's *Die Rakete zu den Planteenraumen* (The Rocket into Interplanetary Space). His imagination was fired and he set about writing a book - *Der Vorstoss in den Weltenraum* (The Advance into Space) – which was published in 1924 and aimed to popularise Oberth's ideas. It was spectacularly successful and by 1930 was in its sixth edition. Valier became a one-man publicity machine for the concept of space travel, publishing articles with titles such as Berlin to New York in one hour and A Daring Trip to Mars.

By the late 1920s Valier was collaborating with Fritz van Opel – of Opel cars fame – to develop a number of rocket-powered cars and aircraft. Whilst van Opel saw these experiments as good publicity for his car manufacturing company, Valier saw them as cementing the concept of rocketry in the public imagination.

Our hero was one of the founding fathers of the *Verein fur Raumschiffahrt* – the Spaceflight Society – which set about developing prototypes of rockets and developing launch techniques. Success came on 25th January 1930 when at the Heylandt plant they carried out the first successful test firing using liquid fuel. On 19th April 1930 Valier drove the first rocket car fuelled by liquid propulsion.

Unfortunately, as we have come to expect with all our inductees, Max suffered a catastrophic blow on 17th May of that year when an alcohol-fuelled rocket which he was working on exploded in his laboratory, the shrapnel from the explosion killing him outright.

All was not lost, however, because his protégé, Arthur Rudolph, went on to develop an improved and safer version of his engine, laying down the foundations for what became modern space rocketry which, in turn, provided the technology which allowed men to land on the moon (if, indeed, men landed on the moon), bringing Valier's vision to reality. Valier is commemorated in South Tyrol for his inventive genius and a number of institutions bear his name to this day.

Truly, a great visionary who was killed by his own invention - Valier is a worthy inductee to our Hall of Fame.

46

EDWARD VERNON
(1684 – 1757)

Life on the ocean wave was a pretty brutal affair in the 18th century – not only did you run the risk of ship-wreck and drowning because of the rudimentary navigational skills and the ramshackle construction of your vessels but if you did engage with the enemy the conflicts were bloody and, to cap it all, provisions were disgusting and were likely to cause your health to decline, if you were lucky. Some significant improvements were made to the matelot's lot thanks to the efforts of our latest inductee into our Hall of Fame – Edward Vernon.

Vernon joined the navy in 1700 and enjoyed what could be best described as a rumbustious career. He was promoted to captain in 1706, taking command of the Rye, and had the good fortune of narrowly avoiding the naval defeat in Sicily in 1707 in which his commander, Sir Cloudesley Shovell – great name – and 2,000 sailors perished.

Vernon played a prominent part in an odd encounter which was known as the War of Jenkins' Ear. Robert Jenkins was a merchant seaman who claimed to have had his ear cut off by Spanish coastguards in the Caribbean. Vernon, who by this time was in parliament, took up Jenkins' case with gusto, urging reprisals and eventually got his way when in 1739 war was declared. Vernon got himself a post as Vice Admiral and led the fleet along with Major General Thomas Wentworth – he of golf course fame. Vernon captured Porto Bello, a Spanish colonial possession (now Panama), a feat for which he was awarded the freedom of the City of London.

However, Vernon's next encounter with the Spanish proved to be disastrous. The assault on Cartegena de Indias in 1741 in what is now Colombia involved the biggest British amphibious attack before the Normandy landings, involving 186 ships and some 27,000 men. Notwithstanding the superiority in numbers – the Brits were pitted against 6 ships and 3,500 men – the one-eyed, one-armed Spanish commander, Blas de Lezo fought tenaciously and forced the Brits to retreat in disgrace to Jamaica and the collapse of Robert Walpole's government.

Vernon's return to England saw him elected as MP for Ipswich and he continued his interest in naval affairs.

His great contribution to the sailor's lot came on 21st August 1740 when Vernon ordered that their customary tot of rum be diluted with water. Because of the foulness of the water, citrus juice, usually in the form of lemon or lime juice, was added to the water to mask the taste. Miraculously, Vernon's sailors became much healthier than their compatriots – because of the protection that the Vitamin C was giving them – and his recipe was adopted by the navy as a whole.

Vernon was accustomed to wearing clothes made out of grogram – a coarse fabric of silk mixed with wool and often stiffened with gum – and was nicknamed Old Grogram or Old Grog. In honour of him, his concoction came to be known as grog and was drunk on a daily basis in the navy until 31st July 1970.

Edward Vernon – for unwittingly improving the sailor's lot, you are a worthy inductee to our Hall of Fame.

47

JOHN WALKER
(1781 – 1859)

The ability to summon up fire at will was vital for man's development and comfort. It was only when we mastered the ability to generate fire that we could cook and keep ourselves warm at night. Rubbing two sticks together was a laborious process and one that was fraught with frustration. Today with our electric and gas cookers and central heating which spring to life at the press of a button, we give little thought to the struggles of our ancestors to get something lit. The only frustration these days is to be found with a smoker whose lighter has packed up or won't work in the wind.

What was revolutionary and made the creation of fire so much easier was the match. Probably, like me, you don't know the origin of this handy device but this is where the latest inductee to our illustrious Hall of Fame, John Walker, comes in. Born in Stockton-on-Tees, Walker was initially apprenticed to a surgeon but having developed an aversion to surgical operations, decided to change tack, studied pharmacy and opened up a chemist and druggist business in his home town around 1818.

Being an inquisitive sort of chap, Walker was fascinated with the problem of how to create fire. At the time there were a number of mixtures which had been discovered which through their chemical reaction would produce a flame. But no sooner had the flame appeared then it extinguished, leaving the user little time to transfer it to whatever he was trying to set alight. What was needed was a way in which the flame produced by a chemical reaction could be transferred to a slow burning substance such as wood. Once this had been cracked, the user would have more control. This was what Walker stimulated his grey cells to solve.

The development of a match came about by chance, as is often the way with discoveries. A piece of wood which had been dipped into some lighting mixture Walker had been preparing caught light when he scraped it against the rough surface of the hearth. This breakthrough led him to create prototype matches, consisting of sticks of cardboard dipped in the lighting mixture. He further refined the match by applying a coating of sulphur to wooden splints about 3 inches long and tipped with a mixture of sulphide of antimony, chlorate of potash and gum, the sulphur essentially transferring the flame to the slower burning wood.

Walker started to sell his invention locally and they were eagerly snapped up. A box of 50 matches would cost you one shilling and with the box came a piece of sandpaper folded in half, through which the match had to be drawn to light it. Walker called his match Congreves in honour of the rocket and artillery inventor, William Congreve.

There were some problems with the Congreves. The chemical reaction could be extreme and flaming balls could burn brightly and then drop onto the floor burning holes in carpets or, even worse, falling onto clothes and setting them alight. The smell of sulphur was off-putting, although the addition of camphor to the mix made the matches less malodorous.

As you would expect with an inductee to our Hall of Fame, despite this revolutionary discovery, Walker didn't profit from it. A couple of years after Walker's matches were launched on the public, Isaac Holden developed, independently he claimed, a sulphur match which he started selling widely. Walker refused to patent his invention, despite being able to demonstrate conclusively that he had come up with the idea before Holden, and made it freely available to anyone. He only received the credit for his brilliance after his death.

John Walker, for inventing the sulphur match and not profiting from it, you are a worthy inductee to our Hall of Fame.

48

HENRY WINSTANLEY
(1644 – 1703)

Sometimes you have just got to put your money where your mouth is and this trait earns Henry Winstanley his induction into our Hall of Fame.

Henry hailed from Saffron Walden in Essex and was known as a bit of a dabbler with a particular fascination for hydraulic and mechanical gadgets. He filled his house in Littlebury with lots of mechanical gadgets of his own design – it was a local attraction and was known as the Essex House of Wonders – and then in the 1690s he decided to capitalise on this fame by opening a Mathematical Water Theatre, known as

Winstanley's Water Works, in Piccadilly. It was a great success combining fireworks, perpetual fountains, automata and fantastic machines – the most famous was the Wonderful Barrel of 1696 which was able to dispense both hot and cold drinks from the same barrel.

Our Henry was also a merchant and invested some of the loot from his commercial enterprises in five ships. Unfortunately, two were wrecked off Eddystone Rocks, an extensive reef in the Plymouth Sound, between Lizard Point and Start Point. Out of pocket, Winstanley demanded of the Admiralty why nothing had been put down to guide ships around this major obstacle. When he was told that it was too dangerous to mark, Henry impetuously said he would build a lighthouse there himself and the Admiralty, not one to look a gift horse in the mouth, gave him permission to do so.

Work on the octagonal tower started on 14th July 1696. The edifice was impressive, fashioned out of Cornish granite and wood with ornamental features and a glass lantern room where the candles would burn to provide the light. The lighthouse was anchored to the adjacent rocks by 12 huge iron stanchions.

Building work was completed in November 1698, but not without a surprising interruption. England and France were at war at the time and a French privateer destroyed the work done to that point and took Winstanley into captivity. However, our hero was released on the orders of Louis XIV with the words, *"We may be at war with England but we are not at war with humanity"*.

During the 5 years that Eddystone operated, no ship was lost. It had cost £7,814 7s 6d and raised tolls at a rate of 1d per ton of £4,721 19s 3d from passing ships.

Naturally and displaying the characteristically flawed judgement of all our inductees, Winstanley had great faith in his construction and professed that he would be happy to be in it during the greatest storm there ever was. He got his wish because he was on the lighthouse making some repairs on 27th November 1703, the night of the great storm. Unfortunately, the elemental forces pitched against his construction were such that the tower was completely destroyed and no trace was ever found of our hero or his five companions.

Henry Winstanley, you are truly a worthy inductee.

49

THE WINSTONS
(Circa 1968 – 1970)

When I was a youth pretty much every live concert had an obligatory drum solo. This was a cue for the other musicians to refuel on their narcotics and for a large proportion of the audience to empty their bladders. For artists in the recording studio, a drum solo strategically positioned can pad out the number sufficiently to give the record buyer the sense of receiving value for money. But occasionally, very occasionally, the rhythm pounded out on the skins has a profound and earth-shaking effect. This is what happened to the latest inductees to our illustrious Hall of Fame, the Winstons.

Based in Washington DC, the Winstons were a 1960s funk and soul group. They were in the studios recording their EP, Color Him Father, in 1969 and were struggling with what to record for the B side. They hit on an instrumental – many a 45rpm single in those days had instrumentals as B sides, showing the group had run out of ideas or material (or both) – which was loosely based on an old gospel called Amen, Brother. But the gospel wasn't long enough and so midway through the number, the drummer, G C Coleman, banged away on his own for four bars and the recording was in the can. The jury is still out as to who master-minded the solo – was it the lead singer, Richard L Spencer, as he claimed or was it the drummer, as the only other surviving member of the band, Phil Tolotta claimed? We will never know.

Anyway the A side was a million-seller but Amen, Brother sank into the obscurity its genesis merited. Notwithstanding their chart success the group, a mixed-race ensemble, struggled to get bookings in the southern states of the United States and broke up in 1970. A not unusual story, certainly unremarkable, but why are the Winstons worthy of our attention?

Well, it is that four bar drum break in the middle of Amen, Brother, of course. If you click on the link and listen to it, there is something very familiar about it: https://www.youtube.com/watch?v=qwQLk7Ncp O4&feature=player_detailpage

Known as the Amen Break it is probably one of the most sampled bits of music ever recorded with several thousand records using it. The beauty of the break is that the rhythm is syncopated so there are lots of variations that can be derived from sampling the original break. It is also sonically very punchy and is both very organic-sounding and robotic. It became the corner stone of genres such as drum and bass, hip hop, jungle, big beat, industrial and electronica.

To illustrate the diversity of artists who have used it, a slightly slowed down version appeared on Salt-N-Pepa's I Desire on their 1986 debut album. The break was used on 3rd Bass' Words of Wisdom and on NWA's 1989 Straight Outta Compton. Even Oasis used it in their 1997 song, D'You Know What I Mean, and David Bowie's hit song, Little Wonder, from his Earthling album features it at the start.

You would expect a snatch of music so influential to have made the Winstons a load of money. But, alas, not so which is why they are inducted into our Hall of Fame. In the 1980s the position of samples and copyright was very sketchy at best – at least nowadays permission is sought - and so the Winstons didn't get a bean.

For your role in launching a whole range of new musical genres, the Winstons, you are worthy inductees into our Hall of Fame.

50

XEROX CORPORATION
(Founded 1906)

There is an old joke that the man who invented the first wheel was an idiot whereas the man who invented the next three was a genius. It seems that this canard may hold true in the corporate world if the experience of the Xerox Corporation, the latest inductees into our illustrious Hall of Fame, is anything to go by.

The name Xerox is synonymous with photocopying and the corporation has made a tidy business out of what they do but it could have been oh so much more as this cautionary tale demonstrates. Xerox had a research department in California called the Palo Alto

Research Company where eggheads were given licence to experiment and develop new devices. Butler Lampson was inspired by the work of another of our inductees, Douglas Engelbart, and in particular his development of the oN-Line System. In a memo written in 1972 he proposed the development of a personal computer, later to be known as the Xerox Alto, which would incorporate all the then developments in computing. Chuck Thacker picked up the ball and was responsible for the design work.

By the following year the work was completed. The Alto boasted an impressive collection of hardware and software features. It had a 606 by 808 monochrome bit-mapped display, a 3 button mouse, large 2.5 megabyte removable disks, a 16-bit microcode programmable CPU and a total address space of 64k 16-bit words including the graphics bitmap. On the software front it had the first WYSIWYG document preparation systems, an email tool, a graphics editor and an early paint program. It even came with one of the first network-based multi-person video games, Gene Ball's Alto Trek. It was truly the world's first PC.

Initially 30 units were manufactured and piloted successfully. A further 2,000 were made and donated to various institutions. But the corporation had no interest in developing the Alto as a commercial proposition. There were probably two reasons for this. Firstly, they were minting money with their 914 copier, the first successful commercial plain paper copier. You couldn't buy them; you had to rent them and pay for every copy you made. Until the patents ran out it was a gold mine and so there was no commercial imperative to look beyond the copying industry. The second reason was more defensive. Some saw a machine that created electronic documents and that was able to transmit them electronically as a threat to their existing paper based business.

But perhaps more surprisingly, they allowed visitors to look around their PARC laboratories. One interested visitor was a certain Steve Jobs – heard of him? – who was given access to everything PARC was up to. He took copious notes and within months had hired some of PARC's brightest sparks and started on development work which produced Lisa, the forerunner of the Mac.

Xerox did develop a PC, the Xerox 820, which incorporated the design outlined in Lampson's memo and included most of the Alto's features. But it was more expensive and for Xerox it was too little too late. The Apple and Sun machines, based on much of Xerox's pioneering work, had cornered the market.

Xerox Corporation, for developing the first personal computer and not realising what you had got, you are worthy inductees into our Hall of Fame.

Printed in Poland
by Amazon Fulfillment
Poland Sp. z o.o., Wrocław